Richard Owen

Memoir on the Megatherium

Or, Giant Ground-Sloth of America

Richard Owen

Memoir on the Megatherium
Or, Giant Ground-Sloth of America

ISBN/EAN: 9783337189839

Printed in Europe, USA, Canada, Australia, Japan

Cover: Foto ©ninafisch / pixelio.de

More available books at **www.hansebooks.com**

MEMOIR

ON THE

MEGATHERIUM,

OR

GIANT-GROUND-SLOTH OF AMERICA

(*Megatherium americanum*, CUVIER).

BY

RICHARD OWEN, F.R.S., D.C.L.,
SUPERINTENDENT OF THE NATURAL HISTORY DEPARTMENTS IN THE BRITISH MUSEUM;
FULLERIAN PROFESSOR OF PHYSIOLOGY IN THE ROYAL INSTITUTION OF GREAT BRITAIN;
FOREIGN ASSOCIATE OF THE INSTITUTE OF FRANCE, ETC.

WILLIAMS AND NORGATE,
14, HENRIETTA STREET, COVENT GARDEN, LONDON;
AND
20, SOUTH FREDERICK STREET, EDINBURGH.
1861.

TO

THE ROYAL SOCIETY OF LONDON,

THIS WORK, which owes its origin to the Society's favourable reception of the Author's communications on the MEGATHERIUM, and allied fossils, in the years 1851 and 1855, and the drawings for its illustration chiefly to the liberal application thereto of part of the 'Government Grant,' is, by permission of the PRESIDENT AND COUNCIL, most gratefully and respectfully dedicated, by

THE AUTHOR.

CONTENTS.

		Page
§ 1.	Historical Introduction	3
§ 2.	Of the Spinal Column	12
§ 3.	Comparison of the Spinal Column	25
§ 4.	Of the Skull	28
§ 5.	Of the Teeth	36
§ 6.	Comparison of the Skull and Dentition	41
§ 7.	Of the Bones of the Anterior Extremity	45
§ 8.	Of the Bones of the Posterior Extremity	62
§ 9.	Comparison of the Bones of the Hind-foot	74
§ 10.	Physiological Summary	77
§ 11.	Geological Summary	82

DESCRIPTION

OF

THE SKELETON OF THE **MEGATHERIUM**.

§ 1. *Historical Introduction.*

BEFORE commencing the description **of the skeleton of** the Megatherium, now in London, Plate I., which is the most complete that has yet reached Europe, a brief statement may be premised of the chief steps which have led to the restoration of the species (*Megatherium Americanum*, CUVIER and BLUMENBACH) to which it belongs.

CUVIER, in communicating to the 'Annales du Muséum' (t. v. **1804**) a translation of the first memoir on this subject—that, viz. by GARRIGA and BRU, published at Madrid in 1796,—gives all the requisite details respecting the discovery of the skeleton therein described, and adds his own more important deductions as to its affinities from an examination and comparison of the plates of the Spanish work.

It appears that proofs of these plates were transmitted in 1795 to the Institute of France, and that CUVIER, having been called upon by the 'Class of Sciences' at that period to give a report upon them, developed his views of the affinity of the animal to the *Sloths* and other *Edentates**, and proposed for it the name 'Megatherium†.'

M. ROUME, the correspondent **of the French Institute to whom that** distinguished scientific body were indebted for the proof impressions of GARRIGA's **work, and** who **had an** opportunity of examining the skeleton itself at Madrid (which CUVIER never **enjoyed**), inserted a brief notice of it in the 'Bulletin de la Société Philomathique' of the Republican year IV. (1795); **in which,** after particularly noticing that the pelvis

---

* " Je développai dès-lors l'affinité de cet animal avec les *paresseux* et les autres *édentés*."—Annales du Muséum, t. v. p. 377.

† μέγας *great*, θηρίον *beast*.

A 2

was open towards the abdomen, the pubis being absent, without any indication of its having ever existed*, concludes that the animal was intermediate, as to form, between the Cape Anteater (*Orycteropus*) and the Great Anteater of America (*Myrmecophaga jubata*). But he adds, that M. Cuvier, from an examination of the engravings of the skeleton, had recognized that the species was much more nearly allied to the Sloths than to the Anteaters.

M. Abildgaard, a professor at Copenhagen, having had the opportunity of studying the skeleton of the Megatherium at Madrid in 1793, published a short notice of it in Danish, illustrated by a rude figure of the skull and of the hind limbs, and referred the species to the *Bruta* of Linnæus, an order afterwards modified to form the *Edentés* of Cuvier: this notice, though published the year after Cuvier's Report, appears to have been independent of it, and the conclusions to be the result of the author's own observations and reflections. It is, therefore, to be regarded as an additional testimony to the true affinities of the species.

Cuvier's comments on the figures in the engravings for Garriga's memoir are accompanied by reduced copies of them, given in the above-cited volume of the 'Annales du Muséum,' and afterwards in the fourth volume of the first edition of the 'Recherches sur les Ossemens Fossiles,' 4to, 1812. In both works Cuvier sums up his conclusions as to the habits and food of the Megatherium, as follows:—"Its teeth prove that it lived on vegetables, and its robust fore-feet armed with sharp claws, make us believe that it was principally their roots which it attacked. Its magnitude and its talons must have given it sufficient means of defence. It was not of swift course, nor was this requisite, the animal needing neither to pursue nor to escape†."

In the year 1821 Drs. Pander and D'Alton published their beautiful Monograph on the Megatherium, the result of personal examination and depiction of the then unique skeleton at Madrid; they represent the bones more artistically and in more natural juxtaposition than in the plates of Bru's memoir, but the subject being the same, the same deficiencies, to be presently specified, were unavoidable.

As the accomplished and learned authors of the German work reasoned, like Abildgaard, from actual inspection of the fossil skeleton, their conclusions as to the nature, affinities and habits of the animal to which it belonged merit a respectful consideration. In it they recognize, with Cuvier, all the important points of resem-

---

* "Son bassin est composé des os sacrum, iléum et ischium, mais il n'y a point de pubis ni d'indication qu'il ait existé. Ce bassin est ouvert du côté de l'abdomen."—p. 97. It will be shown in this memoir that the supposed want of pubic bones was due to accidental mutilation of the pelvis of the skeleton at Madrid: the true profile of the pelvis is given at e_1, e_2, e_3, Plate I.

† "Ses dents prouvent qu'il vivoit de végétaux, et ses pieds de devant, robustes et armées d'ongles tranchans, nous fait croire que c'étoit principalement leurs racines qu'il attaquoit. Sa grandeur et ses griffes devoient lui fournir assez de moyens de défense. Il n'étoit pas prompt à la course, mais cela ne lui étoit pas nécessaire, n'ayant besoin ni de poursuivre ni de fuir." *Tom. cit.* 'Sur le Mégathérium,' p. 29. See also the posthumous edition of the 'Ossemens Fossiles,' 8vo, 1836, tom. viii. p. 363.

blance to the skeletons of the existing species of Sloth, of which they append excellent figures. But, imbued with the principles of the transcendental and transmutative hypotheses, then prevalent in the schools of Germany, they regard the great Megatherium and Megalonyx as being not merely predecessors but progenitors of those still lingering remnants of the tardigrade race, into which such ancestral giants are supposed to have dwindled down by gradual degeneration and alteration of characters. But they deem the living habits of the Megatherium to have been far different from those of its puny scansorial progeny: it was, in their opinion, a fossorial animal; and not merely an occasional digger of the soil, as Cuvier concluded, but altogether a creature of subterranean habits; some earth-whale, as it were, or colossal mole. Pander and D'Alton nevertheless give to this animal, which they truly characterized as one of the most extraordinary of its class, the name of 'Riesen-faulthier,' *Bradypus giganteus*, or Gigantic Sloth[*].

Cuvier, in preparing his new and enlarged edition of the famous 'Recherches sur les Ossemens Fossiles,' availed himself, in the fifth volume, published in 1823, of the labours of the German anatomists and draughtsmen, just cited, and substituted copies of their figures for those which he had previously borrowed from Bru.

The teeth of the Megatherium are still described as being implanted by two roots, and as being sixteen in number, formulized as $m, \frac{4-4}{4-4}$: there is the same deficiency of the sternal ribs, pubic bones and tail; the manubrium sterni continues to be represented in a reversed position: but, with regard to the bones of the fore-foot, the organization of which was involved in obscurity, owing to the faulty manner in which Cuvier believed them to have been articulated, he endeavours to throw some new light on their arrangement. After a comparison of the figures given by Pander and D'Alton with the bones of the fore-foot in existing Edentata, Cuvier concludes that the fore-feet in the Madrid skeleton are transposed, the right being on the left, and the left on the right side; that the index, medius and annular digits were the only ones provided with claws; that the thumb was clawless, and the little finger rudimental and concealed, in the living Megatherium, under the skin; the hand being thereby specially formed for cleaving the soil and digging, like that of the *Dasypus gigas*. On this hypothesis, names are applied by Cuvier to certain bones of the carpus, none of which had before been determined[†]. Cuvier, however, adds, that "in order to verify his conjectures it must be necessary to have access to the skeleton itself, and to compare separately each bone of the fore-foot with their homologues in that species of Armadillo[‡]." His ideas of the affinities of the Megatherium have now undergone some

---

[*] Das Riesen-Faulthier, &c. fol. 1821, p. 16.

[†] Recherches sur les Ossemens Fossiles, 4to. t. v. pt. 1. 1823, p. 185. pl. 16. fig. 13. The letters indicative of the carpal bones have been omitted, by oversight, in the plates, but there is no difficulty in adding them according to the description given by Cuvier in the text.

[‡] "Mais on sent que, pour vérifier ces conjectures, il faudroit être auprès du squelette, et en comparer séparément tous les os avec leurs analogues dans ce tatou, ce que j'espère que quelque anatomiste espagnol ne tardera pas à faire."—*Ib*. p. 185.

modification: the following paragraph is added to the summary on this head given in the earlier edition of the great work:—"Its analogies approximate it to different genera of the Edentate family. It has the head and the shoulder of a Sloth, whilst the legs and the feet offer a singular mixture of characters peculiar to the Anteaters and Armadillos*." CUVIER concludes his account of the Megatherium in the second edition of the 'Ossemens Fossiles,' by appending a note communicated to him by M. AUGUSTE ST. HILAIRE, "which announces," he says, "that the Megatherium had pushed its affinity to the Armadillos so far as to be covered like them with a scaly cuirass."

This opinion derived apparent confirmation from the description by Professor WEISS of portions of an osseous tessellated dermal armour of some gigantic quadruped, sent to Berlin by the traveller SELLOW, which armour he figures, and attributes to the Megatherium, in a "Geological memoir on the Provinces of S. Pedro do Sul and the Banda Oriental," published in 1827 †.

In the year 1832, a highly valuable and important collection of the bones of the Megatherium, discovered in the Rio Salado, with a portion of a bony tessellated dermal covering of an animal, found in Lake Averias, province of Buenos Ayres, indicative of a frame as great as that of the skeleton from the Rio Salado, was transmitted from Buenos Ayres by Sir WOODBINE PARISH, K.H., and presented by him to the Royal College of Surgeons. These specimens formed the subject of a memoir communicated by WILLIAM CLIFT, Esq., F.R.S., to the Geological Society, June 13, 1832‡, in which, although, with the characteristic caution of the author, the armour is not directly affirmed to belong to the Megatherium, nothing is stated to prevent the inference that it formed part of the 'Remains' of that animal which it is the object of the memoir to describe: and, in the description of the map engraved in pl. 43, the specimen figured in pl. 46, with other portions of the bony armour, are comprehended amongst "those Remains of the Megatherium which have hitherto been sent to Europe §." Further countenance to the later opinion of CUVIER as to the affinities of the Megatherium to the Armadillo, was afforded by a few remarks in the text of Mr. CLIFT's memoir: thus, in noticing the "bony or pseudo-cartilaginous pieces which unite the true ribs to the sternum," Mr. CLIFT adds, "as is also the case in the Armadillo ‖." And in the description of the caudal vertebræ, he remarks, "they have the inferior spines (i. e. the chevron or V-shaped bones), manifesting in this their relation to other Edentata, as the *Myrmecophagæ* and *Dasypodæ* ¶."

---

* "See analogies le rapprochent des divers genres de la famille des édentés. Il a la tête et l'épaule d'un paresseux, et ses jambes et ses pieds offrent un singulier mélange de caractères propres aux fourmilliers et aux tatous."—*Ib.* p. 189.

† Abhandlungen der Kön. Akad. der Wissenschaften zu Berlin, 1827.

‡ Geological Transactions, 2nd series, vol. iii. p. 437.

§ *Ibid.* Description of the Plates.   ‖ *Ibid.* p. 439.   ¶ *Ibid.* p. 444.

The additional parts of the Megatherium, supplied by Sir WOODBINE PARISH, and deficient in the skeleton at Madrid, were two of the ossified cartilages of the ribs, two of the smaller bones of the sternum, twelve caudal vertebræ, and ten of the separate 'chevron bones,' partly belonging to them and partly indicating other caudal vertebræ: they also included a part of the os hyoides. Mr. CLIFT, with his accustomed ingenuity and artistic ability, gives at one view an idea of "all the parts hitherto known, or supposed to be known," of the Megatherium, by taking as his basis the outline of the view of the skeleton given by PANDER and D'ALTON in the first plate of their work; leaving in simple outline those parts which are present in the Madrid skeleton, but not in Sir WOODBINE PARISH's collection; expressing by a pale tint the parts in that collection which also exist in the Madrid skeleton; and indicating by a dark tint the additional parts which are deficient in the Madrid skeleton, and had not before been figured.

Besides the important elements thus added towards the completion of our knowledge of the skeleton, Mr. CLIFT was enabled to correct an error into which CUVIER had been **led by a figure of a** mutilated tooth in tab. 4. fig. 5, F, of GARRIGA's memoir, which seemed to show that it had been implanted, as CUVIER describes it to have been, by two fangs. PANDER and D'ALTON give a similar figure of one of the **teeth** in their tab. 2. fig. 15. The figure of the natural size of one of **the teeth of the Me**gatherium transmitted by Sir WOODBINE PARISH, given in the third plate (pl. 45. fig. 2.) of Mr. CLIFT's memoir, **is the first accurate representation** of these characteristic parts, and shows that the implanted base is widely excavated for a persistent matrix, as in the Sloths and Armadillos.

The prevalent belief among Comparative Anatomists and Naturalists at this period, founded upon the additional observations by WEISS and CLIFT to those contained in the second edition of the 'Ossemens Fossiles' of CUVIER, may be gathered from such notices as were then published of the opinions expressed by the eminent professors of those sciences on the subject. Thus Dr. ROBERT GRANT, treating of the Armadillos, in his Lectures on Comparative Anatomy, says, "The Megatherium itself appears to have been such a digging loricated animal, and in many of its bones resembles the Armadillos *."

The Very Rev. Dr. BUCKLAND, in his Bridgewater Treatise published in 1836, admitting the probability, from the evidence at that time adduced, that the Megatherium had been defended by a bony tessellated armour, argues that—" A covering of

* Lecture, reported in the 'Lancet,' March 29, 1834; and again, in 1839, the Megatherium is described by Dr. GRANT as being "allied in structure to the *Bradypus*, and shielded with cutaneous plates like the *Dasypus*." —THOMSON's 'British Annual' for 1839, p. 274. M. DESMAREST, in the art. *Megathère* of the 'Dictionnaire des Sciences Naturelles,' 1823, writes as follows:—" Leurs membres étaient robustes et terminés par cinq gros doigts. Des observations récentes paroissent prouver que sa peau, épaissée et comme ossifiée, était partagée en une foule d'écussons polygones et rapprochés les uns des autres comme les pièces qui entrent dans la composition d'une mosaïque."—" La forme des molaires et la taille de ces animaux semblent indiquer qu'ils se nourrissoient de végétaux et sans doute de racines."

such enormous weight would have been consistent with the general structure of the Megatherium: its columnar hind legs and colossal tail were calculated to give it due support; and the strength of the loins and ribs, being very much greater than in the Elephant, seems to have been necessary for carrying so ponderous a cuirass as that which we suppose to have covered the body." He next calls attention to the broad and rough flattened surface of a part of the crest of the ileum, to the broad summits of the spines of many vertebræ, and also to the superior convex portion of certain ribs, on which the armour could rest, as affording "evidence of pressure, similar to that we find on the analogous parts of the skeleton of the Armadillo, from which," he remarks, "we might have inferred that the Megatherium also was covered with heavy armour, even had no such armour been discovered near bones of this animal in other parts of the same level district of Paraguay*."

The estimable and justly celebrated author of the 'Bridgewater Treatise,' notwithstanding his bias for the hypothesis of the affinities of the Megatherium to the Armadillos, enunciates his conclusion with philosophic caution, and affirms that the other "remarkable character of the Megatherium, in which it approaches most nearly to the Armadillo and Chlamyphorus, consists in its hide having probably been covered with a bony coat of armour, varying from three-fourths of an inch to an inch and a half in thickness†." In the same work is given an original figure of the pelvis and hind limb of the Megatherium, from a front view of those specimens in the Museum of the College of Surgeons.

M. LAURILLARD, in the posthumous edition of the 'Ossemens Fossiles' of CUVIER, published in 1836, whilst admitting it to be very possible for the Megatherium to have been covered by a cuirass, appends a note of warning against too hastily attributing to that animal the fragments of the gigantic osseous armour that had been found in the same formations of South America; because, in the casts of some of the bones which were transmitted with that armour by Sir WOODBINE PARISH, M. LAURILLARD had recognized a calcaneum, an astragalus and a scaphoid, differing from those of the living Armadillos only by their size and by some merely specific modifications‡.

But that which Baron CUVIER and M. LAURILLARD had ventured to regard as very possible, and Dr. BUCKLAND as probable, M. DE BLAINVILLE a few years later announced to be a positive fact. He communicated, in 1839, to the Academy of Sciences of the French Institute, a statement that bones of the Megatherium had recently been discovered, accompanied with fragments of a carapace belonging indubitably to the same animal; and he adds that the association of a bony armour with the internal skeleton of the Megatherium can be demonstrated as surely by à priori reasoning as by the à posteriori fact; but he adduces no observations or

* Geology and Mineralogy considered with reference to Natural Theology, vol. i. pp. 160 and 161.
† Ibid. p. 159.
‡ Recherches sur les Ossemens Fossiles, 8vo, 1836, tom. viii. p. 354.

arguments from the skeleton in addition to those of which Dr. BUCKLAND had previously availed himself, simply affirming that "the Megatherium is proved to have been certainly covered by an osteo-dermal carapace, by the disposition of the spinous processes of the vertebræ, by the angles of the ribs, by the articulation of the pelvis with the vertebral column," &c.; and he concludes by announcing "that the Megatherium was a gigantic species of Armadillo, most nearly allied to the diminutive *Chlamyphorus* *."

With regard to the fossils from South America, unequivocally referable to the Armadillo family, I had myself pointed out the generic distinction of that large quadruped, some bones of which had been transmitted along with the gigantic dermal armour by Sir WOODBINE PARISH, and proposed for it the name of 'Glyptodon' in Sir WOODBINE PARISH's work on Buenos Ayres †; and afterwards, stimulated by the general tendency of anatomists and palæontologists to regard the Megatherium as being, likewise, a gigantic Armadillo, I entered upon a critical review of all the facts of the case which at that time had been obtained, and communicated the result in a memoir to the Geological Society, read March 23, 1839‡. The general conclusions from this memoir were:—

1. The opinions of CUVIER and WEISS, in favour of the Megatherium being so armed, rest on no better ground than the mere fact of bony armour of some gigantic quadruped and the skeleton of the Megatherium having been discovered in the same continent.

2. The skeleton, or its parts, which have been actually associated with the bony armour above mentioned, belongs to a quadruped distinct from and less than the Megatherium.

3. No part of the skeleton of the Megatherium presents those modifications which are related to the support of a dermal covering.

4. The proportions of the component tesseræ of the bony armour in question to the skeleton of the Glyptodon are the same as those between the dermal tesseræ and skeleton of existing Armadillos, but are much smaller as compared with the bones of the Megatherium.

5. No bony armour composed of tesseræ having the same relative size to the bones of the Megatherium as in the Glyptodon and existing Armadillos, has yet been discovered.

In 1837 I had been put in possession of an additional test of the affinities of the Megatherium, by portions of teeth, obtained by Mr. CHARLES DARWIN at Punta Alta in Northern Patagonia, from which specimens I was kindly permitted to take the requisite sections for microscopical examination. Previous researches by Professor

* "Recherches sur l'ancienneté des Edentés terrestres à la surface de la terre," Comptes Rendus de l'Acad. des Sciences, 1839, p. 65.
† P. 178 b, frontispiece, 8vo, 1838.
‡ "On the *Glyptodon clavipes*," Transactions of the Geological Society, second series, vol. vi. p. 98.

Retzius and myself into the structure of the teeth of the Mammalia generally, had made me acquainted with the marked difference between the teeth of the Armadillos and those of the Sloths in internal structure, and I now found that the Megatherium presented the same remarkable compound structure of the teeth as in the Sloths, but with additional complexity, by which they still further departed from the comparatively simple structure of the teeth of the Armadillos: the examination at the same time proved that there was no true enamel in the teeth of the Megatherium*.

Another important evidence of the affinity of the Megatherium to the Sloths was brought to light by a fragment of the skull from Punta Alta, which demonstrated a fifth small molar tooth on each side of the upper jaw, thus showing that in the number as well as in the structure and the kind of teeth the Megatherium agreed with the *Bradypodidæ*, and especially with the Ai or Three-toed Sloth; the anterior pair of molars not manifesting the excess of size and laniary form which characterize them in the Unau or Two-toed species†.

These additional evidences of the concordance of structure between the Megatherium and the Sloths, manifested by the hard and enduring parts which are most intimately related to the food of the animal, induced me to reconsider the conclusions of Cuvier, Pander and D'Alton, and Dr. Buckland relative to the sources of its nutriment and its habits of life; and ultimately to arrive at a conviction of the correspondence of the food of the Megatherium with that of the Sloths, and of the relation of the modified form of the Megatherium to its peculiar mode of obtaining such food, the grounds for which conviction are given in my memoir on the *Mylodon robustus*, published in 1842.

By the analogy of this smaller species of the great extinct terrestrial Sloths of South America, I endeavoured to dissipate some of the doubt and obscurity which shrouded the true structure of the fore and hind feet of the Megatherium‡; but the light so obtained served rather to increase the desire to inspect the skeleton itself at Madrid, and obtain, *ex visu*, a conviction of the accuracy of my views; for I participated entirely in the doubts expressed by my experienced colleague Mr. Clift§ relative to that skeleton, then unique in Europe, viz. as to "whether it had been properly or improperly mounted, *i. e.* whether all the parts were of one or more individuals, whether they belong to the situation or position in which they are placed, whether all the parts are genuine or partly modelled, or whether parts are eked out by bones that do not belong to the part or situation in which they are collected;" concurring at the same time with Mr. Clift, that "no blame was attributable to the articulator, who, probably, had little or no guide in such a difficult task."

---

* Zoology of the Voyage of Her Majesty's Ship 'Beagle,' Fossil Mammalia, 4to, 1838–40, p. 103, pls. 31 and 32. fig. 1.   † Ibib. p. 102.

‡ Description of the Skeleton of the *Mylodon robustus*, 4to, pp. 102, 131–136.

§ "Notice on the Megatherium," Geol. Trans. Second Series, vol. iii., description of plate 44.

Year after year, however, passed away without bringing with it the requisite liberty from official duties for a visit to Madrid. I availed myself of the rare opportunities afforded by journeys of scientific friends to that city to endeavour to obtain information on some of the discrepant or doubtful points, and more especially relative to the exact number of teeth or sockets of teeth in the skull; but usually without any satisfactory result, owing chiefly to the difficulty which the mode of preserving the famous skeleton presents to any close inspection. Dr. DAUBENY, the accomplished Professor of Botany at Oxford, in reply to one of my requests, wrote to me from Madrid:—"I have examined the Megatherium and can discern only four teeth in either jaw, which are all perfect and double, but whether or not there be the rudiment of a tooth behind cannot be distinctly ascertained, unless the glass case which covers the specimen be removed; for there is no door or any way of getting close to the skeleton. I should advise a memorial to be drawn up stating the reasons for wishing the point to be determined, in which case, perhaps, the authorities might consent to allow a pane of glass to be removed."

Fortunately the necessity of the endeavour to overcome these obstacles was in great measure obviated by the arrival in 1845, in this country, of a very important and remarkable accession of remains of the Megatherium, discovered in 1837, near Luxan, Buenos Ayres, which, with other fossils of large extinct South American animals, were purchased by the Trustees of the British Museum. This collection included an entire cranium and lower jaw of the Megatherium; all the vertebræ of the tail; complete series of the bones of both fore- and hind-feet:—in short, parts which, in combination with those previously deposited in the Museum of the Royal College of Surgeons, by Sir WOODBINE PARISH and Mr. DARWIN, completed the skeleton of the animal, including the hyoid and sesamoid bones, which are too often wanting in the skeletons of our recent and common quadrupeds.

Accurate plaster casts of the huge pelvis and most of the other bones of the Megatherium in the College Museum had been prepared at the expense of the College and presented to the British Museum: and, after an examination and comparison of the whole series of the remains of the Megatherium in both collections, and a consultation with the ingenious articulator of the Mylodon, Mr. FLOWER, I suggested to the able keeper of the Mineralogical Department in the British Museum, CHARLES KÖNIG, K.H., F.R.S., the advantage which would arise, if similar plaster casts should be taken of all the other bones of the skeleton, and if such casts, coloured to match the original bones, should be mounted; the originals being preserved separate, for the greater facility of their comparison and for the advantage of examining their articular surfaces.

The Trustees of the British Museum having taken the proposition under mature consideration, the Council of the Royal College of Surgeons having liberally sanctioned the moulding of all the requisite bones in their Museum, and my own consent to superintend the co-adjustment and attitude of the skeleton having been given, the models and plaster casts were ordered by the Trustees to be made, and their articulation was confided to Mr. FLOWER.

The result is the exhibition in our National Museum of the entire skeleton of the Megatherium, Plate I., in a much more complete state, and, I believe I may add, more natural attitude, than that of the same extraordinary quadruped, which previously had been unique and the glory of the Royal Museum of Natural History at Madrid.

For the full fruition by comparative anatomists and palæontologists of so rich an accession to our evidences of one of the strangest animals of a former world, there still remained one condition,—viz. the power to employ a competent artist to depict the skeleton and its several parts. There could be no question that the opportunity of supplying the omissions, correcting the errors, and clearing up the doubts, in the descriptions founded on the Madrid skeleton, ought to be embraced without loss of time. The enlightened and liberal grant of £1000, placed by Lord JOHN RUSSELL, then Prime Minister, at the disposal of the Council of the Royal Society, in aid of the labours of men of science, seemed to me to afford the means of removing the only difficulty that stood in the way of completing the object of my wishes. I therefore submitted the case to the "Committee of Recommendations for the application of the Government Grant," and the Council of the Royal Society was pleased to approve the recommendation of the Committee, viz. "that £100 be granted to Professor OWEN, to be applied to the procurement of drawings of the undescribed and unfigured or inaccurately figured parts of the skeleton of the Megatherium."

The present memoir, and its illustrations from the accurate pencil of Mr. JOSEPH DINKEL, are the result of that recommendation; and I have only to add, that the descriptions and figures of the several parts of the skeleton of the Megatherium, now in London, have been taken from the actual bones; and the views of the entire skeleton from the articulated casts, which are so beautifully exact, as, for all the essential purposes of science, to be of the same value and utility as the bones themselves would be if so articulated together.

§ 2. *Of the Spinal Column.*

The skeleton of the Megatherium, like that of all other vertebrate animals, being composed of a series of segments, similar in their composition, and referable under all their modifications to a common type, answering to that which is figured as the 'typical vertebra' in my work on the Vertebrate Skeleton*, I shall commence its description by one of those segments which deviate least from the archetypal character; and such segments we find to constitute the major part of the trunk, where they form what, in human anatomy, would be termed 'dorsal vertebræ,' 'ribs,' 'cartilages of ribs' and 'sternal bones.'

* On the Archetype and Homologies of the Vertebrate Skeleton. 8vo, 1848, p. 81, fig. 16, and p. 82, fig. 15.

Plate II. fig. 1, gives a front view of the fifth segment of the dorsal or thoracic region of the trunk: it deviates from the archetype inasmuch as the neurapophyses, n, n, have coalesced, as in other mammals, with the centrum, c, below, and are connate with the neural spine, ns, above: the hæmal arch is also vastly expanded in relation to the greatly developed vascular centres which it was destined to encompass; the pleurapophyses, pl, pl, being elongated and bent down, and the hæmapophyses, $h.h$, removed from the centrum and articulated to the ends of the pleurapophyses, and, by a double synovial joint, $s's'$, to the hæmal spine or sternal bone $h.s$.

The coalesced centrum and neural arch constitute the so-called 'dorsal vertebra,' and the one selected is the fifth of that series counting backwards.

The centrum, c, or body of the vertebra, is wedge-shaped, with its base upwards, forming the floor of the capacious neural canal, and the sides—slightly concave lengthwise, almost flattened vertically—converge to the inferior surface, which is formed by an obtuse ridge: the centrum expands slightly at its articular ends, and so that the contour of the anterior one is rather oval than trihedral; this surface is slightly depressed at its middle, slightly convex in the rest of its extent; the posterior articular surface is larger than the anterior one and flatter, but is also a little depressed at the middle: the two upper angles of the hinder end, probably contributed by the neurapophyses in the development of the vertebra, are slightly produced and truncate, offering each a smooth, flat, small subcircular surface, c', for the head of the rib of the preceding vertebra; the corresponding part of the rib of the present segment is marked c''. The neurapophyses, n, n, rise each by a slender base which has coalesced with the anterior half of the upper and outer angle of the centrum: they diverge from each other and expand as they rise; then, developing some articular surfaces and exogenous processes from their outer surface, arch towards each other, increasing rapidly in antero-posteriorly extent, and coalesce above the neural canal; where they support the zygapophyses, z, z, and the thick and strong neural spine ns. The roof of the neural arch, thus formed, projects some way beyond the anterior surface of the centrum, and extends almost to the posterior surface. The inner surface of the neural arch is as remarkable for its even smoothness, as the outer surface is for its various prominences and depressions. The outer side of the basal half of the neurapophysis supports a large elliptical articular surface (a'), concave from above downwards and backwards: the overhanging fore-part of the arch supports the two flat oval anterior zygapophyses, z, z, the articular surfaces of which look almost directly upwards; on the under surface of the back part of the arch are the two posterior zygapophyses looking almost downwards, and between these is a rough longitudinal prominent ridge. The neural spine is moderately long, subcompressed and subtrihedral, with a sharp anterior margin, smooth sides, and a rough thick posterior surface, developing a median longitudinal ridge: the summit expands into a rough triangular almost flattened surface.

For the convenience of describing and comparing the different exogenous processes developed from the neural arch in the class *Mammalia*, I have indicated them by single-worded names; having found that, although they varied much in size and a little in relative position, when traced through the series, they could be identified from species to species.

The most common and constant of these processes is that which usually stands out, or transversely, from the base of the neural arch, and affords a joint or a surface of confluence for the rib; this I have proposed to call '*diapophysis**'; the second process has a range of variety in its position from the upper part of the diapophysis to that of the anterior zygapophysis, but, as it is commonly somewhere between these two, I have called it '*metapophysis*†'; the third process projects most commonly more or less backwards, from the base of the diapophysis, and I have termed it '*anapophysis*‡'. As each of the above processes developes in some species an articular surface, and as each is usually more or less oblique in position, I have called those processes which more constantly support such surfaces '*zygapophyses* §'. The comparative anatomy of these and other 'exogenous' processes has been the subject of a previous memoir ‖, because their extreme degree of variety in the Order *Edentata*, and extraordinary development in the Armadillos and true Anteaters, render them of unusual importance in the question of the affinities of the Megatherium.

The narrow compressed base of the neurapophysis of the fifth dorsal vertebra in that animal having risen above the centrum, developes on its outer surface a large vertically oval articular surface, n', concave in the direction of its long axis, almost flat transversely; to which surface a corresponding convex articular surface, n'', on the upper part of the neck of the rib near the head, is adapted. Above the surface, n', the diapophysis, d, stands out, short, thick, subdepressed, expanded at its extremity, which is slightly produced backwards and supports on its outer surface an elliptical articular concavity, d', with the long axis directed from above downwards and backwards, and articulated to a corresponding convexity, d'', upon the tubercle of the rib. A rugged tuberosity, m, on the upper and fore-part of the diapophysis represents in rudiment the metapophysis. A smaller tuberosity is interposed between the anterior zygapophysis, z, and the neurapophysial surface, n', for the rib. Thus the neural arch of the vertebra in question presents ten distinct articulations besides the two sutural ones now obliterated by its anchylosis to the centrum, which part has its two large terminal articulations distinct from those for the head of the rib, c', which I reckon among the neurapophysial ones. The rib, which term I confine to the

---

* Lectures on the Comparative Anatomy of the Vertebrate Animals, 8vo, Longmans, 1846, vol. i. p. 42; from διά *trans*, and ἀπόφυσις *processus*.

† On the Anatomy of the Male Aurochs, Proceedings of the Zoological Society, November 14, 1848, p. 131; from μετά *inter*, and ἀπόφυσις.

‡ On the Anatomy of the Aurochs, *ut supra*, p. 131; from ἀνά *retro*, and ἀπόφυσις.

§ Lectures on the Comparative Anatomy of the Vertebrate Animals, p. 43; from ζυγὸν *junctura*, and ἀπόφυσις; these are called the "oblique or articular processes" in Human Anatomy.

‖ Philosophical Transactions, 1851.

pleurapophysis, or 'vertebral rib' of comparative anatomists, 'pars ossea costæ' of anthropotomy, presents a small flat subcircular surface, c'', for articulation with that on the base of the neurapophysis forming the upper angle of the body of the vertebra in advance of the segment to which the rib belongs. The neck of the rib rapidly expands as it quits the head, developes the convex oval surface, n'', on its upper part for the concavity on the neurapophysis of its own vertebra; and is indented behind where it joins the tubercle. The oval convex articular surface, d'', for that on the diapophysis, is situated on the upper and towards the back part of the tubercle. The body of the rib is moderately convex on its outer side, more convex transversely on its inner side, where the convexity is bounded by a groove on each side, extending half-way down the rib, near its rather sharp margins: in its lower half the rib becomes a little broader and less thick. The outer surface of the rib is well marked by grooves and ridges for muscular attachment; the best-developed eminence being at a short distance from the tubercle; and the largest and deepest groove being behind the tubercle.

The hæmapophysis, or 'sternal rib,' h, is a straight subcompressed bone, with a very irregular surface, which is somewhat convex on the inner side, but is traversed by strong oblique ridges with intervening deep and wide channels on part of the outer side. The surface of junction with the pleurapophysis is a very rough and irregular one for ligamentous union: the opposite end of the hæmapophysis divides into two convex condyles, $s's'$, separated by an oblique, deep and rather narrow groove; the outer condyle projects further than the other; on the shorter one the articular surface passes continuously from one side to the other, describing a semicircle; on the longer condyle the articulation is divided by a median constriction into two oval convex surfaces. One half of each of the condyles articulates with a corresponding concavity on the 'sternal bone' (hs) of its own segment, the other half of each condyle with the contiguous sternal bone.

The sterneber or sternal bone, completing, as 'hæmal spine,' hs, the typical segment in question, is a cuboid piece, divided into an outer or peripheral, and an inner or central portion. The outer portion is subpentagonal, having its four corners excavated by as many concavities for the hæmapophysis; the two upper concavities, $s''s''$, being divided by a flat rough tract, the two lower concavities by a rough tuberosity. The outer surface is flat and rough. The inner portion, or that next the cavity of the chest, is larger than the other and has a hexagonal contour; the four angular concave articular surfaces, s'', s'', for the hæmapophyses being separated, at the sides of the bone, by rough tracts; and, above and below, by a flat articular surface, hs, by which the bone articulates with contiguous sternal bones. The inner surface of this portion is flat and rough, having been apparently covered by a strong aponeurosis in the living animal. Thus the whole bone presents not fewer than ten articular surfaces, viz. a flat semicircular one above or in front of, and a similar one behind, the posterior division; and two concave articulations, s'', s'', on each side of both divisions for the double condyles of two pairs of hæmapophyses.

The chief modification in the sixth segment of the chest is the development of a third articular surface at the back part of the base of the spine, between the two posterior zygapophyses; and the somewhat greater production of the ridge which stands out from the fore-part of the base of the spine between the anterior zygapophyses. The inferior or hæmal arch is also augmented by increased length of the ribs. The key-bone of that arch, sterneber or hæmal spine, Plate XI. figs. 4–7, repeats the characters of that of the previous segment, save that the sternal articulation, s, and the hæmapophysial ones, hp, hp, of the central division of the bone are more continuous, as shown in fig. 6. Fig. 5 shows the surface which was presented towards the integument; fig. 4 that which was turned towards the cavity of the chest; fig. 7 is a side view showing the four articular cavities for the double condyles of the sixth and seventh hæmapophyses; fig. 6 shows the under surface of this remarkable type of sternal bone.

In the seventh dorsal vertebra (Plate III. figs. 1, 2, 3), a third articular surface, mz, fig. 1, is developed between the two anterior zygapophyses, z, z, to join that upon the back part of the sixth vertebra; so that there are three zygapophyses, a median and two lateral, on both the fore and the back part (mz', fig. 2) of the arch of this vertebra, making, with the three articular surfaces on each side (fig. 3, c', n', d') for the ribs, and with the anterior and posterior surfaces of the centrum, not fewer than fourteen joints. In this and the two following segments of the back (D_8 and $_9$, Plate I.), the ribs attain their greatest length. In the tenth segment the hæmapophyses cease to articulate below directly with a hæmal spine. The median zygapophyses continue to be developed both before and behind to the twelfth dorsal vertebra inclusive. In the thirteenth (Plate X. fig. 4) this supplementary articulation is suppressed behind; and the costal articulations have disappeared from the diapophyses d. Those on the neurapophyses are almost circular, n', and those on the upper and posterior angles of the centrum, c', have increased in size. The metapophysis (m), which was indicated by a protuberance above the diapophysis in the preceding dorsals, begins to assume the form of a rugged thick vertical ridge. The fourteenth dorsal vertebra shows the progressive increase of size of the centrum, and the absence of the median zygapophysis before as well as behind; and in it the costal articulation on the centrum for the penultimate rib is lost, as well as that on the diapophysis for the antepenultimate one, and only the subcircular concave neural surface for the rib remains. In the fifteenth dorsal vertebra the posterior zygapophyses are convex transversely at their inner border, slightly concave in the rest of their extent; the back part of the neural spine between these processes is deeply grooved; the metapophysial-ridge increases in height and length; a short and thick anapophysis is developed from the back part of the base of the diapophysis, and on the under part of the anapophysis there is a distinct, nearly flat, articular surface. The sixteenth dorsal vertebra (Plate III. figs. 4 and 5, Plate X. fig. 5) offers a corresponding modification at the fore-part of each neurapophysis, in the development of a short, strong, wedge-shaped process, p, fig. 5, answering to the parapophysis in *Myrmecophaga** and *Dasypus*, with an articular surface, *pa* (Plate III. fig. 4), on its upper part for junction with the anapophysis of the preceding vertebra. The metapophysis, *m*, projects forward above the zygapophysis, the articular surface of which (*z*, fig. 4) is continued upward upon the metapophysis (*mz*, fig. 4). The anapophysis, *a a*, fig. 5, Plate X., forms a strong thick square plate of bone projecting upward, outward and backward from the diapophysis. The anapophysial articular surface, a', a', fig. 5, Plate III., is on the under and back part of this plate, nearly parallel with the posterior zygapophysis, z', the convex inner border of which has increased in thickness. The body of the vertebra assumes, with its larger size, a more decided trihedral or wedge-shaped figure at this part of the spine.

The rib of the thirteenth segment loses the convex articular surface on the tubercle, which is attached by ligaments to the diapophysis. The rib of the fourteenth segment loses the small flat surface at the extremity of the head; and only the large convex surface on the upper part of that end of the neck remains, which surface, extending to the free end of the neck, reduces that part to an edge: the tubercle exists in this rib, and is rough for ligamentous insertions as in the preceding. In the fifteenth (Plate III. figs. 1 and 3) and sixteenth (*ib.* fig. 2) ribs the tubercle subsides; the neck of the rib is defined by the rugosity of its whole upper surface, save that part where the articular convexity, n'', remains for articulating with the neural arch, as shown at n', fig. 5, Plate III.

The hæmal spine ('sterneber') of the eighth segment, Plate XI. figs. 8–12, may be the last of the so-called bones of the sternum. It is divided, like those in advance, into a peripheral and a central portion. The peripheral portion (fig. 9) is of a subquadrate form, the four corresponding articular surfaces (*ha*, *ha'*) for the hæmapophyses almost touching each other at their margins; the outer roughened surface is convex; the anterior hæmapophysial articular surface is suppressed on the left side of this division of the bone, fig. 10, to which the corresponding hæmapophysis seems to have been united by ligament. The central division of the bone (fig. 8) presents the median flat surface (*s*, fig. 10) on its upper or fore-part for the antecedent sterneber, and the concave hæmapophysial surface, *hp*, *hp*, on each side of its anterior half; but posteriorly the hæmapophysial surface on each side is confluent with that of the same side belonging to the peripheral division of the bone, which thus presents at its lower or hinder part only two long oval concave articular surfaces (*hp a*, figs. 11 & 12) for the pair of hæmapophyses of the ninth segment of the chest, and the concavities almost meet at the under surface of the sterneber, which there presents no articular surface for a succeeding one. The number of articular surfaces, therefore, of this bone is reduced to six; one, *s*, for the antecedent sterneber, two on the anterior half of the right side, *hp*, *ha*, fig. 12, for the bifid condyle of the eighth hæmapophysis, one on the anterior half of the left side, *ha*, fig. 10, for one of the condyles of the opposite hæmapophysis; and a pair of surfaces on the posterior and lateral parts, *hp a*,

* Philosophical Transactions, 1851, Plate L. fig. 22, *p*, *pa*.

for the ninth pair of hæmapophyses which terminate each by a single convex condyle. It is possible that a more simplified sterneber may have intervened between the hæmapophyses of the tenth segment.

In the three segments of the trunk, Plate I. $L_{1,2,3}$, succeeding the last of the dorsal series, both pleurapophyses and hæmapophyses are wanting as distinct ossified parts, and those segments are reduced to the coalesced elements, constituting the 'lumbar vertebræ' of Human Anatomy. The accessory articulations between the parapophyses and anapophyses are continued in these vertebræ, which do not become anchylosed together in the Megatherium as in the Mylodon.

I next proceed to trace the modifications of the segments as they recede from the typical one in the opposite direction or towards the head.

The fourth dorsal segment much resembles the fifth, which has been taken as the type; the pleurapophyses are shorter, especially at their cervix; but the complex articulations of these and of the hæmapophyses are repeated.

In the third segment the pleurapophysis, pl, fig. 3, Plate IX., and hæmapophysis, ib. h, are anchylosed together: both are shortened, but the pleurapophysis in a greater degree; this retains its three articular surfaces, c'', n'', d'', on the head, neck and tubercle; and the hæmapophysis, h, has its double condyle, s', s'', at the sternal end, fig. 3 b, the inner one being single, the outer one divided by a narrow groove into an anterior and a posterior convexity. The concave border of the rib is less produced than in the fifth segment.

In the second dorsal segment (**Plate I. D_2**) **the neural spine** is increased in height, **and the metapophysial tubercle is diminished in size.** The vertebral, pl, and sternal, h, parts of the rib (Plate IX. fig. 2) are anchylosed, and both are shortened: the former retains its **three** articular surfaces on the head, c'', neck, n'', and tubercle, d'', that on the tubercle being the largest, and being partially divided into two convexities (fig. 2a, d). The pleurapophysis (pl, fig. 2) is diminished in length, but increases in breadth to its place of coalescence with the hæmapophysis (h); the convex articular surface (fig. 2b, s'') on the outer condyle of the hæmapophysis is not divided; the inner condyle, s', is much reduced and has only a small articular surface.

The first dorsal segment (Plate I. D_1) is remarkable for the superior height and antero-posterior extent of the neural spine, the summit of which expands into a broad flat subtriangular rough surface, Plate IV. fig. 5. D_1, ns. The anterior margin of the spine is sharp and produced. The anterior zygapophyses are not so near each other as in the succeeding vertebræ; and they are continued outwardly upon the base of the metapophysis, which is here more distinct from the diapophysis than in the succeeding vertebræ; and the articular surfaces of the zygapophyses are slightly concave transversely. The costal concavity below the diapophysis is continuous with the smaller articular surface upon the side of the neurapophysis.

The pleurapophysis (Plate IX. pl, fig. 1) is much **reduced in** length, and is confluent below with a short, thick, broad, subquadrate hæmapophysis, h. The short neck of the rib terminates in a small obtuse end without any distinct articular surface: this end seems to have been imbedded in the ligamentous mass between the seventh cervical and first dorsal vertebræ: the short neck quickly expands into the shaft of the rib: a small elliptical surface, fig. 1 a, n'', on the upper part of the neck, is continued at its outer end into the larger convex surface upon the upper part of the tubercle, d'': from the middle of the anterior border of this surface a strong ridge is continued down the outer surface of the rib to its hinder border. The sternal end of the hæmapophysis, fig. 1 b, presents a large subtriangular surface, slightly concave in one direction, slightly convex in the other, adapted to a similar single concavo-convex surface on the side of the much developed and modified hæmal spine called 'manubrium sterni.'

This bone (Plate XI. figs. 1, 2, 3) is of an oval or cordiform figure, with a prominence on each side near its inferior truncated apex, below the lateral articulations, hp, for the first pair of sternal ribs. The outer surface is principally concave lengthwise, and is sculptured by the impressions of the coarse aponeurotic structures which have been adherent to it in the living animal: a short median longitudinal ridge projects from its lower part. The inner surface is chiefly convex, but is very irregular. At its upper half a strong median prominence divides the shallow rough depressions, cl, for the attachment of the clavicular ligaments: these depressions are deepest above the costal articulations which are supported on well-marked triangular prominences. Between these prominences the bone is rather concave. A strong rough tuberosity projects below the lower angle of the costal surface. The contracted inferior end of the manubrium terminates in an oval convex articular surface, s', for the second sternal bone. There are no articular surfaces for the hæmapophyses of the second dorsal segment.

The skeletal segment (Plates I. C_7; IV. fig. 7) in advance of the first of the dorsal series is reduced to its centrum and neural arch; both pleurapophysis, hæmapophysis and hæmal spine are absent; and its remaining anchylosed elements constitute the 'seventh cervical vertebra' of Descriptive Anatomy. This is most remarkable for the great development of the neural spine (Plate IV. fig. 7, ns), which exceeds that of the first dorsal vertebra: the summit is similarly expanded and flattened above (Plate IV. fig. 5, ns 7); but the anterior margin is still more produced, forming a low angle about half-way down the spine (fig. 7, ns). The anterior zygapophyses are more remote from each other than in the first dorsal, and their articular surfaces are chiefly supported by the inner side of the base of the metapophyses, figs. 5, 7 m, which are here well developed and more distinct from the diapophyses, d, fig. 7, than in the dorsal region. The diapophyses, figs. 5 & 7, d, are strong, rugged, stand out from the sides of the neural arch, and terminate in rough truncate ends. The posterior zygapophyses (figs. 6 & 7, z') are slightly convex.

In the sixth cervical vertebra (Plates I. C_6; V. figs. 5 & 6) the spine, ns, is much shortened, though still long and large in proportion to the neural arch. The metapophyses, m, m, stand out from the upper part of the side of the neural arch behind the anterior zygapophyses, z. The diapophyses, d, d, are developed from the base of the neural arch, which has descended lower upon the sides of the centrum; and now we find another element—the pleurapophysis, pl—restored to the segment, but reduced to rudimental proportions and anchylosed at two points. Its vertebral end is bifid; one portion, answering to the head of the rib, has coalesced with the side of the centrum (at p, fig. 5); the other, answering to the tubercle, has united with the under part of the diapophysis, d: what may be termed the body of the rib is a short but broad rhomboidal plate (fig. 6, pl), projecting outward, downward and a little backward. The space intercepted between the pleurapophysis and diapophysis forms the canal, v, for the vertebral artery.

The fifth cervical (Plate I. C_5) differs from the sixth by its smaller dimensions, especially by its shorter spine, and by the diminution in the breadth of the pleurapophysis, which terminates by a thick obtuse end: it sends out, however, a thin plate forwards from its vertebral end. The metapophysis is a large obtuse tubercle, Plate IV. fig. 5, m 5.

In the fourth and third cervicals (Plate V. figs. 3 & 4) the neural spine is still more reduced, and, contracting from its base, assumes a triangular shape, fig. 4, ns. The anterior zygapophyses, fig. 3, z, are concave transversely, and look upward and inward; the posterior ones, fig. 4, z', are convex, with the reverse aspect: the metapophysis, m, continues to be developed as a distinct tuberosity, external and posterior to the prozygapophyses; and the pleurapophysis continues to send forward the pointed plate from its fore-part, fig. 4, pl, its outer end, pl, being thick and tuberous, like the diapophysis, d, above.

The dentata (Plate V. figs. 1 & 2) has its spine extended in the antero-posterior direction, and of great strength, though low; with a thick angular ridge projecting from its fore-part and overhanging the neural canal; it is broad, flattened and almost vertical behind, and has a subbifid summit (Plate IV. fig. 5, z, and Plate V. fig. 1, ns). There are no metapophyses and no anterior zygapophyses, but the analogous articular surfaces (figs. 41 & 42, zn) have descended upon the antero-lateral parts of the coalesced centrum of the atlas or 'odontoid process,' and are adapted to corresponding surfaces of the bases of the neural arch of the atlas. The posterior zygapophyses, z', are wide apart, and are convex. The diapophysis, d, is short and obtuse; the pleurapophysis, pl, still developes its anterior angle, Plate V. fig. 2, pl'. The fore-part of the odontoid process, o, is a rounded tuberosity, on the under surface of which is the oval, slightly convex surface for articulating with the hypapophysis (Plate IV. fig. 3, o, hy), which has coalesced with the neural arch, ns, of the atlas, and is commonly called the 'body of the atlas.' The under surface of the centrum of the dentata developes a hypapophysial ridge.

The atlas, viewed from behind, as in Plate IV. fig. 3, is a large, transversely oblong, subdepressed, shuttle-shaped bone, perforated by a large aperture, quadrate below for its detached centrum the 'odontoid process,' arched above for the spinal cord.

The thinnest and smallest part of the ring of the atlas is formed by the hypapophysis, *hy*, which has coalesced with part of the bases of the neural arch, *m*, and has supplanted, as it were, the proper centrum, *o*, Plate V. figs. 1 & 2, which has remained anchylosed to that of the axis. The upper surface of the hypapophysis presents a shallow articular surface, *o*, Plate IV. fig. 3, for that centrum to rest and turn upon. The hinder half of the base of the neurapophysis developes, on each side, a slightly concave, subcircular, articular surface, *z'*, for a moveable articulation with that on the side of the odontoid, *zn*, fig. 1, Plate V. The atlas is perforated anterior and external to this by a foramen, Plate IV. fig. 1, *s*, answering to that called 'foramen alare posterius' in the Horse, in which it gives passage to the posterior branch of the occipital artery; in the Megatherium the foramen or canal is bridged over by a narrow oblique bar of bone, dividing its external outlet into two, *r* & *s*, and through the hinder, *s*, of the divisions it is probable that a branch of the second spinal nerve may have passed.

The diapophysis, *d*, is a broad, depressed, rounded aliform process, with a protuberance from its under and back part, like the rudiment of a pleurapophysis, *pl*. Anterior to this process the under surface of the diapophysis is deeply and widely excavated and perforated by the vertebral artery, the canal for which, opening upon the upper surface of the diapophysis, is then continued obliquely inward, perforating at *q*, fig. 4, Plate IV. the upper part of the neural arch, just within the upper part of the condyloid concavities. A large part of the canal for the first spinal nerve, fig. 1, *r*, opens into the outer commencement of the vertebral canal, and answers to that called 'foramen alare anterius' in the Horse, which transmits the inferior branch of the first spinal nerve as well as the anterior branch of the occipital artery. The condyloid concavities, fig. 2, *nz*, are semioval, large and deep, and occupy nearly the whole of the anterior surface of the neural arch, being separated above by a rough tract of three inches' extent, upon which the vertebral canals open. There is a triangular rough surface at the back and inner part of each condyloid concavity.

Such are the modifications of the different cervical vertebræ of the Megatherium. With regard to the dorsal vertebræ, their chief characteristics may be briefly recapitulated as follows:—

The first dorsal vertebra is distinguished by the confluence of the neuro-costal and dia-costal surfaces, and by the superior height of the spine.

The second to the fifth dorsals inclusive, like the first, have only the ordinary pair of zygapophyses before and behind, but have the neural and diapophysial surfaces for the rib distinct.

The sixth dorsal is recognizable by having a third median zygapophysis behind, but not in front. The seventh to the twelfth dorsals inclusive have the three zygapophyses both before and behind. The thirteenth dorsal has the median zygapophysis in front but not behind: the costal surface has disappeared from the diapophysis. The fourteenth dorsal has only the ordinary pair of zygapophyses before and behind, but may be distinguished from the second, third, fourth and fifth dorsals by the absence of the costal articulation on the diapophysis, and of that on the upper and hinder angle of the centrum. The fifteenth dorsal has an anapophysis on each side with an articular surface, and has only the costal articulation on the neurapophysis. The sixteenth dorsal has on each side, at the fore part of the neural arch, a parapophysis with a superior articular surface, and behind, an anapophysis with an inferior articular surface. But it differs from the lumbar vertebræ by the costal surface on the neurapophysis.

Having now described the principal characters of those segments of the skeleton, the centrums and neural arches of which are comprehended in Anthropotomy under the term of 'true vertebræ,' on account of their freedom of motion on each other, I next proceed to the description of the 'false vertebræ;' and first, of those that, being anchylosed together, form the 'sacrum.'

This part of the skeleton includes five vertebræ (Plate VII. 1–5), which are not only anchylosed to each other, but to both the iliac and ischial bones: the length of the sacrum is 22 inches, its extreme breadth across the fifth vertebra, fig. 1, d_5, is 20 inches. The centrum of the first vertebra (Plate VI. c) presents a transversely oblong, subquadrate, flattened, articular surface for that of the last lumbar vertebra, with its margin a little produced forwards, and developed below into a pair of rough ridges, * *. The neural arch overhangs this surface, and developes a metapophysis from the fore part of each side of its base, with a broad articular surface on its under part, and a similar surface above (Plate VII. fig. 1, z), representing the anterior zygapophysis; the two surfaces meeting at a right angle at their inner borders. The broad diapophysis of the first sacral, 1, is perforated by a small subvertical canal, d', at its confluence with the ilium, and is separated from the corresponding part of the next diapophysis by a larger orifice, o_1, which is the first of the four superior or posterior sacral outlets. The neural arch of the first sacral vertebra, n_1, is separated from that of the second, n_2, by a narrow transversely elongated elliptical vacuity. The neural arches of the three succeeding vertebræ are completely confluent: a pair of triangular closely approximated apertures, n_4, divides the base of the neural spine of the fourth, ns_4, from that, ns_5, of the fifth sacral vertebra. The neural spines of the first four sacral vertebræ have coalesced into a strong vertical ridge, ns, ns_4, increasing in thickness as it extends backwards, and being there from two-thirds of an inch to one inch and a half thick across the broken summit. The second posterior sacral canal, o_2, intervenes between the diapophysis of the second and that of the third vertebra. The metapophysis, m_4, of the fourth appears as a low angular tubercle above and a little behind the diapophysis of the third sacral, 3. The diapophysis of the fourth sacral, d_4, extends outwards beyond the ilium, as a subdepressed broad process with a rough free extremity; the back part of the process coalesces with a similar but stronger and longer diapophysis of the fifth sacral, $d\,s$, from the fore-part of the base of which a tuberous metapophysis, $m\,s$, projects upward and forward. The third, $o\,s$, and fourth, $o\,q$, upper sacral outlets are wider apart than the second and first; the sacrum expanding posteriorly. The back and under part of the diapophysis of both the fourth and fifth vertebræ coalesce with the ischium and with the thick and strong parapophysis extended from the side of the centrums. The neural arch of the fifth sacral developes a pair of posterior zygapophyses, Plate VII. z', z', with a flat surface looking outward and a little downward, and with the lower angle continued upon a small rough subarticular surface. The posterior surface of the last sacral vertebra, Plate VII. fig. 2, is on the same vertical parallel as the posterior zygapophyses; it is nearly flat and transversely elliptic. The neural canal of the sacrum, the anterior aperture of which is 3 inches in vertical and 4 inches in transverse diameter, expands in the sacrum, and opens below by three wide foramina on each side: of these the first and second are of great size: into the second foramen the third upper sacral canal leads: the third lower sacral foramen, which is the smallest, corresponds with the fourth upper one: the fifth canal for the fifth pair of sacral nerves broadly grooves the back part of the parapophysis and side of the centrum. The posterior aperture of the neural canal is 2 inches in vertical and 4 inches 3 lines in transverse diameter. Both diapophyses and parapophyses of the first three sacral vertebræ coalesce with the ilia. The sacrum is concave below both transversely and lengthwise.

The tail of the Megatherium was of great strength: it is so long as to touch the ground when the trunk is raised at an angle of forty-five degrees from the horizontal position: it includes eighteen vertebræ, which progressively diminish in size from the first to the last, Plate I. $C\,d$, 1–18.

The first vertebra, Plate II. fig. 2, is remarkable for the length and strength of its diapophyses, d, which are expanded at both ends, and, like those of the sacral vertebræ, are probably lengthened out by connate or coalesced pleurapophyses, pl. The base of the process, dp, extends from the side of the centrum to the base of the neural arch, is widely excavated behind for the passage of the first pair of caudal nerves, and is subcompressed before it expands into its rugged free termination. These processes are shorter than those of the last sacral vertebra. The neural canal, n, is triangular, 3 inches in vertical and 3 inches 9 lines in transverse diameter. The neural arch developes two posterior zygapophyses, z', with their articular surfaces looking downwards and outwards: two anterior zygapophyses, z, with their articular surfaces looking upwards and inwards, these being strengthened by a strong tuberous metapophysis, m, on their outer side: the spine, ns, is of moderate length, carinate behind, obtuse and slightly expanded above. From the under part of the transversely elliptical centrum are developed two hypapophyses, hy, each with an oblong articular surface, Plate X. fig. 6, hy, to which is joined a hæmapophysis, Plate II. h.

Each hæmapophysis is a long slender conical **bone**, with an articular surface at each angle of the base, hy', hy'', and an obtuse slightly inflected apex: the inner side of the bone is slightly **concave, the** outer one convex transversely: a rough tuberosity dividing it from the inflected apex. The essential differences between the first caudal segment and the dorsal one delineated on the same **Plate** are, that the pleurapophysis, *pl*, is short and anchylosed to the diapophysis, the **hæmopophyses**, *h*, articulate with the centrum, and the hæmal spine is absent, in fig. 2.

The second caudal **vertebra** (Plate VIII. figs. 1 and 2) differs from the first in having an anterior, fig. 2, *hy*, as well as a posterior, ib. hy', pair of hypapophyses; and in the confluence of the hæmapophyses, fig.1, *h*, at their apices forming the so-called 'chevron bone' (*os en chevron*, CUVIER). This vertebra is smaller than the first **caudal** in all its parts except the hæmapophyses, and in all its dimensions except the vertical diameter, which is due to the development of the coalesced parts of those elements into a long and strong hæmal spine, *hs*. The anterior hypapophyses, fig. 2, *hy*, which are the smallest, articulate with the surface on the back part of the base of the hæmapophyses of the first caudal **vertebra: the** posterior hypapophyses, hy', which are more oblong and closer together, articulate with the anterior and larger pair of surfaces, hy', of their own hæmapophyses, fig. 1, *h*. There is a strong rough tuberosity projecting backwards external to each of the anterior hypapophyses. The posterior articular surface is, in the present instance, developed only on the right hæmapophysis, fig. 1, hy'; on the left it is represented by a rough tubercle.

From the third to the fifth caudal vertebræ inclusive, the proximal end of each hæmapophysis has both the large anterior transversely oblong surface for its own centrum, and the smaller subcircular posterior surface for the next centrum: the spine, or coalesced portions, of the third pair is the longest in the caudal series; beyond this it progressively diminishes. In the sixth caudal vertebra the hæmapophyses have a rough **protuberance instead of** the posterior articular surface. After the eighth the protuberance subsides to a rough ridge. In the eleventh caudal (Plate VIII. figs. 3–6) the distal end of the coalesced and shortened hæmapophyses, *hy*, is truncate, and as broad as the divided bases. The under surface of the corresponding centrums of the sixth to the eleventh caudal offers the articular surface on the anterior **pair** of hypapophyses, fig. 6, hy'; the posterior pair, ib. *hy*, are **rough** tuberosities. The posterior zygapophyses (figs. 3, 4, 5, z' z') retain their articular **surfaces to the** tenth caudal: in the eleventh they are mere angular projections. The metapophyses, *m m*, are continued **to the** fourteenth caudal. The neural spine is reduced to a low tuberosity on the thirteenth caudal (**fig. 7**, *ns*); the neural arch continues **complete to the sixteenth, fig. 8,** *n*; in the seventeenth, fig. 10, the neurapophyses, *n*, are mere **exogenous** ridges, bounding the sides of an open neural groove. The **hæmapophyses are** continued **to the** fourteenth vertebra: two pair of rough low hypapophysial tubercles, *hy*, *hy*, fig. 9, continue to be developed to the sixteenth: they subside on the penultimate **caudal**, fig. 11, in which the diapophyses are represented by an obtuse ridge on each side of the centrum. In the last centrum, figs. 12, 13, all the processes have disappeared, and it presents the form of a low rounded cone, with a smooth concave pentangular base, fig. 13, and a rough tuberous summit, fig. 12. This simplified modification of the central element terminates the vertebral series.

§ 3. *Comparison of the Spinal Column with that of other* Bruta.

In the number of the true vertebræ, and of those in each kind, the *Myrmecophaga jubata*, amongst the *Bruta*, agrees with the Megatherium. The Ai, or Three-toed Sloth, *Bradypus tridactylus*, has the same number of dorsal and lumbar vertebræ, but has two more in the cervical region; the Unau, or Two-toed Sloth, *Cholœpus didactylus*, has the same number of cervical and lumbar, but has eight additional dorsal vertebræ, being the greatest number known in any mammalian quadruped. The Short-tailed Manis (*Manis brevicaudata*) has seven cervical and sixteen dorsal vertebræ, but it differs from the Megatherium in having five lumbar vertebræ. The Armadillo tribe (*Dasypodidæ*) differ most from the Megatherium in the inferior number of the dorsal vertebræ, which do not exceed eleven in some species, nor twelve in any. The Orycterope, *Orycteropus capensis*, shows its affinity to the Armadillos in having but thirteen dorsal vertebræ: and, like them, it has five lumbar vertebræ. With regard to the structure of the vertebræ, the Anteaters, both hairy (*Myrmecophaga*) and scaled (*Manis*) most resemble the Megatherium in the length and the uniform backward inclination of the spinous processes; but these processes are not so long in proportion to their antero-posterior extent. The spinous processes of the dorsal vertebræ are short and, in the hinder ones, obsolete in the Sloths: the Unau shows the nearest affinity to the Megatherium by having a few of the anterior dorsal spines better developed than in the Ai. In the Orycterope the last dorsal spine is vertical, indicating a centre of motion in the trunk, those behind and those before slightly converging towards this centre: nothing like this appears in the Megatherium.

In the development of the accessory articular surfaces upon both anapophysis and parapophysis of the last dorsal and lumbar vertebræ, the Megatherium manifests a more direct departure from the Sloths and a proportionate affinity to the Anteaters. The Armadillos, which likewise possess these accessory joints, have superadded peculiarities of the posterior dorsal and lumbar vertebræ, in relation to the support of their peculiar bony armour, of which the Megatherium offers as little trace as do the *Myrmecophagæ*: I allude to the progressively and rapidly increasing length of the metapophysis[*]. These, in the lumbar region, equal in length the spinous process itself; to which the metapophyses bear the same relation in the support of the over-arched carapace that the tie-bearers do to the king-post in the architecture of a roof. From the fact of the metapophyses in the dorsal and lumbar vertebræ of the Megatherium, Plate I. and Plate III. figs. 4 and 5, *m*, not being developed beyond the state

[*] See Plate XLIX. figs. 18 & 19, *n*, Philosophical Transactions, 1851.

of a tubercle, I long ago drew the inference that, like the Sloths and Anteaters, it was not covered by a bony armour*.

With regard to the cervical vertebræ, the fact of the Megatherium having the normal number in the Mammalian class, seven—if it were not sufficiently established by the well-adjusted articulations of those in the skeleton here described, rendering any supplemental vertebræ inadmissible,—would have been made most probable by the same number being present in the skeleton of the Megatherium at Madrid, and in the more complete skeleton of the *Mylodon* in the Museum of the Royal College of Surgeons in London. Moreover, that one of the Megatherioids had seven cervical vertebræ and no more is certain: the skeleton of the *Scelidotherium*, discovered and deposited by Mr. Darwin in the Museum of the Royal College of Surgeons, having been imbedded, without disturbance of the true vertebræ, and those of the neck being exposed in the ordinary number, and in their natural juxtaposition, on the removal of the stony matrix†.

The atlas in both the Ai and Unau presents but two perforations on each side upon the upper surface; one in front, the other behind the base of the transverse process, and this is less produced and is of a quadrate rather than a triangular form.

The dentata of the Unau resembles more that of the Megatherium in the size of the spinous process than that of the Ai does; but the spine in the Unau is pointed behind, not bifurcate. In the forms and proportions of the spines of the succeeding cervical vertebræ the Unau approaches nearer to the Megatherium than does any other existing Edentate species; but the spine of the seventh cervical is by no means proportionally so developed, and metapophyses are not present in any. The Armadillos are distinguished from all other *Bruta* by the great breadth, the shortness and the anchylosis of the middle cervical vertebræ. In the Anteaters (*Myrmecophaga*) the spine of the dentata is low and is extended more forwards than backwards; the spines of the other cervical vertebræ are still less elevated. In the long-tailed Manis very similar proportions of the cervical spines prevail.

The closest correspondence with the Megatherium in the form and structure of the cervical vertebræ is presented as might be expected by its extinct congeners, the *Mylodon* and *Scelidotherium*.

A resemblance of the Armadillos to the Megatherium has been pointed out in the ossification of the sternal ribs, but this is a character common to the order Edentata, and is consequently equally manifested by the Sloths. The Anteaters most resemble the Megatherium in the double joints by which the sternal ribs articulate with the sternum. There is, however, a character by which the Sloths peculiarly resemble the Megatherium, viz. in the anchylosis of the sternal with the vertebral portion of the rib in those of the first three dorsal segments. The Unau most resembles the Megatherium in the form of the manubrium sterni, having the same prolongation of that

* Geological Transactions, 2nd Series, vol. vi. p. 101 (1839).
† Description of the Fossil Mammalia collected during the Voyage of the Beagle, 4to. 1838–40. pl. 20. p. 84.

bone in advance of the expanded part giving articulation to the first rib. This prolongation, which is not present in the *Bradypus tridactylus*, relates to the complete development of the clavicles in the *Cholœpus didactylus*, and serves, as in the Megatherium, for their ligamentous attachment. In other existing *Bruta* the manubrium sterni has a broader and shorter figure, and is generally emarginate anteriorly. The succeeding sternal bones present the nearest resemblance to those of the Megatherium in the genus *Myrmecophaga*, in which they are divided into two parts, each having articular surfaces for those on the bifurcate ends of the ossified sternal ribs; but here the broad depressed portion of the sternal bone is external, the stumpy cylindrical part internal or toward the thoracic cavity. The small subcubical sternebers of the Sloths represent the peripheral divisions only of the more complex bones in the Megatherium. Only the Sloths and Anteaters resemble the Megatherium in the small number of the lumbar vertebræ, and the Megatherium most resembles the *Myrmecophaga* in the number and complexity of the articulations between these. The genera *Manis*, *Dasypus* and *Orycteropus* have five lumbar vertebræ.

Both the *Mylodon* and *Scelidotherium* have three lumbar vertebræ; but these had coalesced with each other and with the sacrum in the skeleton of the *Mylodon robustus* described by me*.

The *Myrmecophagœ* have five sacral vertebræ as in the Megatherium: the Orycteropus has six: so likewise has the *Bradypus tridactylus*: the *Bradypus didactylus* has seven sacral vertebræ; the Armadillos depart furthest from the Megatherium in the unusual number of vertebræ which coalesce to form the sacrum, these ranging from eight to ten in the different species.

It would be unsafe, however, to infer that the Megatherioids were more nearly allied to the Anteaters than to the Sloths in respect of the structure of the sacrum of the Megatherium, for the *Mylodon robustus* has seven sacral vertebræ, like the *Bradypus didactylus*. The posterior expansion of the sacrum and its junction with the ischia, so as to circumscribe the great ischiatic notch, is a character common to the Order *Bruta*. The sacrum early anchyloses with the iliac bones in the Sloths; and the broad and expanded ilia of these animals, with the long and slender anterior parts of the ischia and pubes, circumscribing a large obturator foramen, and bounding a capacious pelvis in front by a narrow symphysis, are characters by which the Sloths, of all existing *Edentata*, offer most resemblance in their pelvis to the Megatherium.

The part of the skeleton in which the Sloths differ most from the Megatherium is the tail, which is so short as to be hardly visible in the entire animal; whilst the Anteaters and Armadillos present the extensive and complex development of caudal vertebræ which characterizes the Megatherioid skeletons. In all the existing species of the ground-dwelling Edentate families the tail is relatively longer and more slender than in the Megatherium; it is even prehensile in the *Manis longicaudatus* and in the *Myrmecophaga didactyla*; but of the modifications of the terminal vertebræ on which that power depends there is no trace in the Megatherium any more than in the Mylodon.

* Description of the Skeleton of the *Mylodon robustus*, 4to. 1842.

The hæmapophyses are distinct, but short and stumpy, in the first caudal vertebra of the *Dasypus longicaudatus*, Pr. Max.: in the *Myrmecophaga jubata* they present proportions much more nearly resembling those in the first caudal vertebra of the Megatherium, and they are equally disjoined at their distal ends *.

The Mylodon in the number, as well as the proportions and structure of the caudal vertebræ, makes the nearest approach to the Megatherium; the hæmapophyses are equally distinct from each other in the first caudal †.

Upon the whole, our deductions from the characters of the parts of the skeleton described in the present section of this memoir, would lead to an inference that *Megatherium* was nearer akin to *Myrmecophaga* than to *Bradypus*; nevertheless the cervical vertebræ, the condition of the anterior ribs, the form of the manubrium sterni, and the pelvis, illustrate the intermediate nature of the giant's affinities, and afford an additional instance to many others which I have had occasion to point out, of a closer adherence to a common type in extinct animals than in the existing species to which they may be most nearly allied.

§ 4. *Of the Skull.*

The skull of the Megatherium is chiefly remarkable for its small relative size, especially in respect of breadth, for the slight diminution of this diameter from the occiput to the end of the nasal bones, for the length and strength of the ascending and descending processes from the orbital end of the zygomatic arch, and for the peculiar depth of the lower jaw, especially at its middle part, where it lodges the long molar teeth.

In the skulls and portions of skulls which have come under my observation, all the cranial and most of the facial sutures, save those that unite the tympanic to the rest of the temporal bone, had been obliterated, and the originally complex assemblage of bones forming the cranium and upper jaw had been reduced to a continuous whole.

Viewed from behind, the skull is remarkable for the degree to which the pterygoid, palatine and maxillary portions descend below the true 'basis cranii': in consequence of which, the foramen magnum and occipital condyles appear to be situated in the upper half of the direct back-view, Plate XIV. fig. 1.

The basioccipital (1) is a broad depressed plate with a thin smoothly rounded concave posterior border: it slightly increases in thickness as it advances forwards to blend with the basisphenoid (Plate XV. 3), but developes on each side a rough protuberance for muscular attachments, anterior to the occipital condyles. In a side view, Plate XII., these condyles form the most prominent parts of the occipital region, which, as it rises above them, slopes forward, giving a very low character of intelligence to the cranium.

---

* See Philosophical Transactions, 1851, **Plate LIII. fig. 60.**

† Memoir on the Mylodon, 4to. p. 69. pl. 8, **fig. 5,** *b, b.*

Each condyle (Plate XIV. fig. 1, *i*, *s*) is a convexity of a subtriangular form, with the base straight and the sides curving to an obtuse apex: the extent of their convex curvature in the vertical or antero-posterior direction (Plate XII. fig. 1, *s*) equals that of a semicircle, and indicates that the Megatherium possessed considerable freedom and extent of motion of the head. The parallel bases of the condyles are turned toward each other and flank the sides of the foramen magnum, Plate XIV. fig. 1, *o*. This foramen is a wide or full ellipse with the longer axis transverse, and the plane, through the greater prominence of the upper border, inclined forward and downward at an angle of 135° with the basioccipital. The plane of the occiput forms, with that of the basis cranii, an angle of 105°. The occipital region (*ib.* 1, *s*, *s*, *s*) is semicircular, its contour being formed by a thick rugged ridge continued from the superoccipital (*s*) on each side to the mastoid, *s*: the osseous wall included between this arched ridge and the condyles (*s*), is divided into four shallow depressions, by a sharp median vertical ridge terminating below at a venous foramen just above the upper border of the foramen magnum, and by a pair of thick obtuse ridges (*r*) extending from the sides of the arched ridge obliquely inwards to near the apices of the occipital condyles, and thence to the paroccipitals (*t*). The spaces so defined are roughened by muscular impressions.

The paroccipital (Plate XIV. fig. 1, *t*) is a moderately developed triangular tuberosity abutting against the back part of the articular cavity for the stylohyal (Plate XV. *ss*), and separated by the outer occipital depression from the mastoid (*s*). The precondyloid canals (ib. *p*), which transmit the motory-lingual or ninth pair of nerves, begin by a large oblique aperture at the middle of the inner side of the base of the condyles and extend along a course of 2½ inches in extent forward and outward to open upon the back part of the rough fossa between the paroccipital (*t*) and petrosal (*st*) externally, and the basioccipital and basisphenoid internally, which fossa answers to the 'foramen lacerum in basi cranii' of ordinary mammals. The carotid foramina (*c*) open upon the fore-part of this fossa. The external precondyloid foramina (*p, p*) are situated on the inner side of the paroccipitals, are each an inch in diameter, and thus indicate the considerable muscular development of the tongue, and its great use in stripping off the leaves and smaller branches of the trees affording the nourishment of the Megatherium.

The basisphenoid (Plate XV. *s*) presents on each of its hinder angles a low rugged subcircular protuberance (*x*) for muscular insertions; in advance of which it slightly expands, the lateral margins inclining downwards so as to form the beginning of the smooth channel that contracts and deepens as it advances towards the posterior aperture of the bony nostrils. The broad subvertical pterygoid plates (*u*) are continued insensibly from the deflected margins of the basisphenoid; they are smooth on the inside, irregularly channelled for muscular attachments on the outside. The posterior aperture of the bony nasal canal (Plate XIV. fig. 1, *n*) is a long and narrow vertical oval, bounded above by a low transverse ridge (Plate XV. *r*), which extends from the hinder border of the bony palate about two inches backwards upon

E

the smooth surface of the pterygoid plates, where it gradually subsides: this ridge forms the angular termination of the bony palate (*ib. u*) behind. The substance of the basisphenoid is excavated by air-sinuses, continued backward from the ordinary position of the sphenoidal sinuses, and extending into the basioccipital as far back as the jugular foramina, *j*. They are exposed by fracture of the walls in the protuberances marked *x, x*, in Plate XV.

From the state of the sutures already alluded to, the limits of the superoccipital, parietal, and frontal bones cannot be defined. Above the arched superoccipital ridge (Plate XIII. fig. 2, *s*) there are two semielliptical rough depressions (*a, a*) for muscular attachments; and, in advance of these, the upper surface of the cranium shows two other similar but shallower muscular impressions, *b, b*. The smooth surface of the parietal, gradually narrowed to an inch in width (*r*) between the temporal ridges (*t*), again as gradually expands into the frontal region (*n*), and it is perforated, a little anterior to the middle of the temporal fossa, by a submedian vascular (venous?) foramen (*v*), about half an inch in diameter. The temporal fossæ (Plates XII. & XIII. *v, t, n*) are remarkable for their great antero-posterior extent, and for the encroachment upon them by the peculiar process (*c*) sent upward and backward from the malar bone, *æ*. They are partially defined anteriorly by the extension of a postorbital process (Plate XII. fig. 1, *n*, Plate XIII. fig. 2, *n*) downward to the malar bone (*æ*); but, beneath and within this slender process, they communicate freely with the orbits, *o*. The posterior boundary ridge, continued from that on the parietal bone, curves forward below and is continued into the sharp upper border of the zygoma, Plate XII. fig. 1, *v*. The surface of the temporal fossæ is grooved and perforated posteriorly by large vessels, and is everywhere strongly impressed by the attachments of muscular fasciculi. The base of the zygomatic process of the temporal bone has an extensive origin (Plate XIII. fig. 2, *v*), not less than 6 inches in antero-posterior extent; its free portion, to where it joins the malar, being 3 inches in length. It is a strong trihedral bar of bone, rather concave on the upper and outer sides, and forming on the underside the glenoid articular cavity for the lower jaw. This cavity (Plate XV. *g*), of an oval form with the long axis transverse, measures 3 inches by 2⅔ inches, and is half an inch in depth. Behind this cavity the base of the zygoma (Plate XII. fig. 1, *v*) has coalesced with the mastoid (*s*) and petrosal (*u*) elements of the temporal, which combine to form the meatus auditorius externus, *e*. This canal is subcircular, about 10 lines in diameter at the deeper part, where it is formed by the above elements, but doubtless wider at its outer part, where it was completed by the tympanic bone. This bone is wanting in the skulls of the Megatherium hitherto transmitted to England: the absence of any fractured surface upon the contour of the orifice of the auditory canal indicates, however, that the bone was a free element of the temporal in the Megatherium as in the Mylodon[*] and Glossotherium[†].

---

[*] Description of the Skeleton of an Extinct Gigantic Sloth (*Mylodon robustus*, Owen), 4to, 1842, p. 28.
[†] Zoology of the Voyage of the Beagle, 'Fossil Mammalia,' 4to, 1840, p. 59. pl. 16. fig. 4.

The mastoid (Plates XII. & XIII. fig. 1, s) forms a rugged process, in depth or length not exceeding the paroccipital (4), but of greater breadth and thickness; above it, externally, and probably in the line of the primitive suture with the squamosal, is a venous foramen, Plate XII. fig. 1, v. The petromastoid—probably the petrosal part in a greater degree—forms the hemispheric articular cavity (Plate XV. 16) for the stylohyal (Plate I. w), the anterior rugged wall of which cavity extends downwards farther than any other part of the proper basis cranii, Plate XII. fig. 1, 14: the petrosal, anterior to this, sends down a shorter rough pyramidal process. The carotid foramen (Plate XV. c), a full ellipse with diameters of 5 lines and 4 lines, is situated between the petrosal and basisphenoid at the fore-part of that oblong depression which is terminated behind by the large precondyloid foramen.

The stylohyal (Plate I. w) has the form of a hammer, with a long, slightly bent handle, terminated by an obliquely truncated rough surface for syndesmosis with the ceratohyal. At the opposite end the handle is subcompressed, and the head is formed by a sudden expansion in the vertical direction, terminated posteriorly by a straight but rugged margin, and with the upper end produced, thickened, and forming a smooth convexity, or condyle, adapted to the cavity above described in the petromastoid. The lower end of the head or expanded part of the hammer-shaped bone is more produced, more rugged, and terminates obtusely. The outer surface has a wide depression at the middle, which is rough, with several short and well-marked ridges. The length of the specimen described is 8 inches, the breadth or depth of the expanded end is 3 inches and a half.

The upper part of the coalesced frontals (Plate XIII. fig. 2, a) forms a smooth triangular plate, rapidly expanding to the postorbital processes (w) and very slightly convex. Some indistinct traces of the fronto-nasal suture seem to show that the nasal bones (Plate XIII. fig. 2, s) extended backward beyond the transverse parallel of the postorbital processes: more distinct traces of the naso-maxillary sutures (n), show that the coalesced nasals were 2 inches 9 lines across at their narrow posterior part, where they are flat above: at first slightly contracting, they then gradually expand, and become more and more convex transversely to their anterior extremity. Here the nasal bones are also thickened, are rugged for the firmer attachment of the cartilaginous parts of the nose, and their under surface, being excavated by two longitudinal grooves, the thickened terminal surface is divided into a middle (Plate XIV. fig. 2, m) and two lateral (n, n) parts, the latter being convex and subangular, and the middle expansion slightly excavated. As in the Two-toed Sloth (*Cholœpus didactylus*), the under surface of each nasal bone sends off a terminal plate or process for the attachment of a turbinal cartilage or ossicle. A narrow median groove indicates the original suture between the nasal bones along their anterior half.

The cranial cavity of the Megatherium is considerably smaller than the cranial part of the skull, the outer wall or plate of bone being separated by large irregular air-cells from the vitreous plate, or that case of bone which immediately invested the brain and its membranes. The vertical diameter of the cranial cavity is 4 inches 8 lines; its transverse diameter, which is greatest at the posterior third part of the cavity, corresponding with the posterior part of the cerebrum, is 6 inches. The brain of the Megatherium, to judge from its bony case, must have been less, by nearly one half, than that of the Elephant; but with the cerebellum relatively larger and situated more posteriorly to the cerebral hemispheres: whence it may be inferred that the Megatherium was a beast of less intelligence, and with the command of fewer resources, or less varied instincts, than the Elephant.

The 'maxilla superior,' or maxillary bone, may be divided into a palatal, alveolar, and facial portion: the latter (Plate XII. fig. 1, n) is remarkable for the excess of its vertical over its antero-posterior extent: it forms, with the coalesced lacrymal (l), the anterior and part of the inferior boundary of the orbit by a strong sub-vertical outstanding plate, curved with the convexity forward, perforated at the middle part of its base by the antorbital canal (r), which is double on the left side, and near the upper part of its thick obtuse margin by the lacrymal canal (l): it is smooth behind, or next the orbit, rather rough and irregular in front: a rough, shallow depression (Plate XIV. fig. 2, s) near the upper part of this surface indicates the origin of a strong labial muscle. The outer surface of the facial plate of the maxillary is smooth and slightly undulated; it evidently extends as far as the postorbital process upwards and backwards, in connexion with the nasal bone: its anterior border (Plate XII. fig. 1, n), terminating the side of the nostril, is vertical, slightly concave and **sharp**, and is smoothly excavated on the inner side or towards the nasal cavity. The lower part of this nasal wall presents a deep and rough sutural notch for articulation with the premaxillary bone.

The alveolar part of the maxillary (Plate XV. i, v) extends about an inch below the suborbital process. The extent of the alveolar tract is 10 inches; its greatest breadth is 2 inches 4 lines, viz. between the second and third teeth. The number of alveoli is five. The first (i) has a subtriangular transverse section, with the apex very obtusely rounded off and turned forward; the borders of this alveolus are sharp and somewhat produced below the level of the surrounding bone. The second alveolus (ii) is close to the first, and the corresponding teeth are nearly in contact; its transverse section is quadrate, the hinder side being the broadest, the outer side the narrowest; the fore-side is more curved than the back one. The partition between this and the third alveolus is thicker than the preceding one, and the teeth stand further apart. The third and fourth sockets are most nearly of a square form, but the transverse diameter predominates; the fifth socket (v) is suddenly reduced in size, and resembles most the first in form, but with the rounded apex of the triangle turned backwards.

No trace of the suture between the maxillary (n) and palatine (w) bones remains: the alveolar border beyond the fifth socket (v) rapidly contracts to the thin vertical pterygoid plate (u).

The bony palate terminates behind in an angular notch, formed by the ridges (*r, r*) before described. The bony palate forms a narrow tract, with parallel lateral borders gently diverging at the fore and back part of the tract, which is very slightly concave transversely: it is perforated by numerous foramina; two long ones, like fissures (Plate XV. *v, v*), opposite the interspace between the third and fourth molars, seem to represent the post-palatal foramina; there are, also, some large foramina (*u, u*) between the first alveoli. The extent of the palatal part of the maxillary in advance of these alveoli is about 1 inch to the hindmost part of the premaxillary (*n*), and 2½ inches to the apex of the process (*n*) articulating with that bone.

The premaxillaries (Plates XV., XII. fig. 1, and XIII. figs. 1 & 2, *n*) have coalesced along the major part of their extent, leaving only a median fissure on their upper surface (Plate XIII. fig. 2, *n*), of about 1½ inch in length, at about the same distance from their slightly expanded anterior ends; at their under surface (Plate XV.) the same fissure is more advanced, and contracts to a few foramina. They form a slender, elongated, subdepressed, four-sided portion of bone, and constitute a singular anterior termination of the skull.

At the base or back part this portion of bone measures 4½ inches across; the fore-end is 2 inches 9 lines across; the narrowest part, near this end, is 2 inches 4 lines across; the vertical diameter is pretty nearly throughout 1 inch 6 lines, but decreases anteriorly. The posterior third of the bone sends upward from its median line a ridge, which enlarges as it approaches the corresponding ridge from the maxillaries, and there presents a smooth and gradually expanding groove at its upper part, for the support of the vomer or its cartilaginous septal prolongation (Plate XIV. fig. 2). Anterior to the median ridge begins the groove which sinks into the fissure, and is then again continued forward as a groove to within an inch of the fore-end of the bone: this part (Plate XV *n*) seems crossed by a rough plate or cap of bone, flat, and about an inch in breadth at its upper part, and there terminating behind, as it does below, in a free margin.

The under surface of the premaxillary mass (Plate XV. *n*) is rather convex antero-posteriorly, as also transversely along its middle third: the groove indicating the primitive suture runs along the whole of this surface, and sinking into its fore-part, opens by two or three foramina into the fissure which is seen on the upper surface. The back part of the under surface of each premaxillary is notched to receive a triangular process of the palatine part of the maxillary (*n*): the more slender median parts of the notches partly divide the prepalatal or incisive fissure (*s*), which thus presents the form of a chevron.

The malar (Plate XII. fig. 1, *x*) is a singularly developed mass of bone, and has always attracted attention as one of the most remarkable features of the skull, from the period of the earliest notices of the Megatherium. Its bulk and complex shape appear to relate to the unusual share which a modified and largely developed masseter muscle must have taken in the act of mastication.

Firmly articulated by extensive reciprocally indented sutures, at one end with the maxillary (ʜ), at the other end with the zygomatic bones (v), and giving an extensive surface of attachment, by a peculiar upward prolongation, to fasciculi of the temporal muscle, it afforded the requisite fixity for the origins of the large and complex masseter.

The suture with the maxillary is in great part obliterated in the skull under description; but a portion remaining on both sides shows that the malar ascended to the level of the antorbital foramen, Plate XIV. fig. 2, r: it forms the lower and a great part of the hinder boundary of the orbit; the latter by a triangular, slightly bent postorbital process (Plate XII. fig. 1, a), which almost touches the corresponding more slender process of the frontal (ib. ɴ). The ascending process (ib. c) is a long, narrow, unequal-sided triangle with an obtuse apex; the descending process (d) is a longer and stronger one, extending, when the mouth is shut, outside and for three inches below the alveolar border of the lower jaw: its extremity is obtuse and recurved. The fourth process (ib. b), which may be called the 'zygomatic' one, extends beneath the end of the corresponding process of the temporal bone, but the obliteration of the suture in the present skull prevents a precise definition of its limits. The whole outer surface of the malar is slightly convex, moderately smooth, with a defined surface for muscular attachment near the back part of the base of the descending process. The inner surface shows, by its well-marked ridges and depressions, the vigorous action of the muscular fasciculi which derived their origin from that part.

The orbit (Plate XII. fig. 1, o), of proportionally small size, as in all large mammalian quadrupeds, presents a long vertically oval form; or rather, by the convex border of the malar (a), is reniform. Its peripheral contour is almost completed by the descending postorbital process of the frontal (ib. ɴ) in the present skull; anterior to which the prominent boundary is effaced by a broad smooth channel, where the orbital surface is more directly continued upon the facial surface of the maxillary: this part answers to the supraciliary notch in quadrupeds. The lacrymal bone being completely coalescent, if not connate, with the maxillary, is recognisable only by the lacrymal foramen (ib. l), which is just within or behind the obtuse anterior border of the orbit. Admitting the essential presence of the lacrymal by this character, it then combines with the frontal, maxillary and malar bones, to form the contour of the orbit. Within this frame, the orbit, as already remarked, communicates extensively with the temporal fossa.

The anterior aperture of the bony nasal canal (Plate XIV. fig. 2, $m, n, \text{ɴ}$) is subcircular, and is formed by the nasals, maxillaries and premaxillaries; the deep vertical sides being contributed wholly by the maxillaries.

The formation of the external bony aperture of the organ of hearing has already been described.

Mandible.—The chief characteristic of the mandible or lower jaw is the near equality of its vertical to its horizontal or longitudinal extent, due to the height of the coronoid process, and more especially to the depth of the dentigerous part of the bone. The latter dimension relates to the interesting modification of the principle of maintenance of the efficiency of the masticating machinery, as contrasted with that in the great proboscidian quadrupeds with a similar diet to the Megatherium.

The condyle of the jaw (Plate XVI. fig. 1, *a*) is transversely elliptical, 3 inches in the long, 1 inch 10 lines in the short, diameter: it is moderately convex, and least so from behind forwards: it seems but a small surface for the articulation of so massive a bone, laden with large teeth, to the cranium; but the adequate firmness of suspension was afforded by the enormous muscles which seem to have embraced every other part of the ascending ramus of the mandible. The coronoid process (Plates XII. and XVI. fig. 2, *b*) was lofty compared with its antero-posterior diameter: it is mutilated in the present skull, but seems to be entire in that of the skeleton at Madrid; and its form and extent may be appreciated in the figures published by Bru* and Pander†. It is much compressed, begins to curve upward immediately anterior to the neck of the condyle, being continued from the middle of that part. The angular process (ib. *c*) of the lower jaw curves backward 4½ inches below the condyle: it is a broad triangular plate, moderately convex externally, concave internally and chiefly by a slight inward bending of the lower margin, Plate XVI. fig. 2, *c*. A few ridges on the comparatively smooth outer surface indicate the insertions of muscles; but the inner surface is strongly sculptured by pits and grooves with strong intervening bony crests. The oblique beginning of the dentary canal (*e*) is situated 6 inches below the condyle, and the foramen is 2 inches from the last alveolus, but above its level. The anterior border of the base of the coronoid process is below the interspace between the fourth and fifth alveolus; on its inner side is a large elliptical outlet of a division of the dentary canal, Plate XVI. figs. 1 & 2, *f*. The outer and inner surfaces of the coronoid process present characters analogous to those on the same surfaces of the angular process, in regard to muscular traces, but the concavity is on the outer side of the coronoid.

The lower contour line of the mandible, which is usually continued forward, straight, or with a gentle curve or undulation, in ordinary quadrupeds, is interrupted in the Megatherium about one foot from the apex of the angular process by a notch, from which the contour line describes an abrupt deep convex curve below the molar teeth, *d*, and then as suddenly rises and passes by a concave curve to the under side of the long and slender symphysis, Plate XII. fig. 2, *d, d*. The depth of the dentigerous part of the horizontal ramus is 9 inches 6 lines; it is slightly convex externally, and forms a flat deep vertical wall internally, ib. *d, iii*.

* Garriga, J. Descripcion del esqueleto de un quadrupedo muy corpulento y raro, &c. fol. Madrid, 1796, Lam. i. and ii. fig. 1.
† Das Riesen-Faulthier (*Bradypus giganteus*), fol. trans., Bonn, 1821, tab. iii. See also Cuvier, 'Ossemens Fossiles,' 4to, tom. v. part 1. pl. 16. figs. 1 and 2.

The antero-posterior extent of the alveolar border is 9 inches (Plate XVI. fig. 1). The first socket is irregularly four-sided, the front side being the shortest, slightly convex, and with the angles rounded off between it and the outer and inner sides: the outer side forms rather an acute angle with the hinder side. The area of the second socket is more regularly quadrate, with the transverse diameter the longest; that of the third socket is nearly a true square; the fourth and last is similar to, but smaller than, the first, with the shortest and most curved side at the back part, and with the antero-posterior diameter a little exceeding the transverse one. The intervals between the alveoli are narrow and subequal.

The rami of the jaw are blended together at the symphysis, which is of great extent, Plate XII. fig. 2, Plate XVI. fig. 1, d,d: it begins posteriorly at the fore-part of the mandibular convexity, opposite the second alveolus, whence the symphysis rapidly contracts to the shape of a scoop or spout, which is prolonged 8½ inches from the alveolar part, and terminates in a thick, rough, rounded and emarginate extremity: the canal at the upper part of this spout-like symphysis is semicylindrical, slightly bent down at the end, and 3 inches in diameter; it becomes roughened by numerous small vascular impressions near the end, but elsewhere is smooth, and has obviously served for the support, during acts of protrusion and retraction, of a long cylindrical tongue. The margins of the canal are thick and rounded. The 'mental foramina,' or anterior outlets of the dental canal (Plate XVI. fig. 2, g), are two on the right side and three on the left, from 4 lines to 8 lines in diameter.

§ 5. *Of the Teeth.*

The teeth are of one kind, molars, five on each side of the upper jaw (Plate XV. and Plate XVII. fig. 2, i, ii, iii, iv, v), four on each side of the lower jaw (Plate XVI. fig. 1, Plate XII. fig. 2, i, ii, iii, iv), eighteen in total number.

In the upper jaw, the first or anterior molar (i) is the second in point of size, the last (v) being the least. The first molar (Plate XVII. fig. 2, i) is 8½ inches in length; the pulp-cavity extends six inches from the base; it presents two slight curvatures, one having the convexity turned forward, and the other inward. The transverse section (Plate XV. i) gives an irregular semicircle, with the convexity turned forward, and the flat side next the second tooth; the angles at which this side joins the curve are rounded; the outer angle is somewhat produced, and the outer side of the curve is flattened. The central axis of the tooth, formed by the vaso-dentine [*], is irregularly tetragonal; the cement is thick on the anterior and posterior surfaces, thin on the sides of the tooth.

The second molar (Plate XV., Plate XVII. fig. 1 & fig. 2, ii) is the largest of the upper series; it exceeds 9 inches in length, is of a tetragonal form, with two slight curvatures, as in the first molar. The posterior and broadest side is nearly flat, the

---

[*] See my 'Odontography,' and Art. Tɛɛᴛʜ in 'Cyclopædia of Anatomy,' vol. iv. for the definition of the different dental tissues.

anterior side somewhat convex, the outer and narrowest side is concave, the inner side is sinuous, having a median longitudinal eminence between two longitudinal concavities. The central axis of vaso-dentine (Plate XVII. fig. 2, v) is more compressed from before backwards than in the preceding tooth, and its posterior surface is concave: the two transverse ridges of the grinding surface of the tooth formed by the dentine (ib. d, d) are nearly equal; but the sloping side formed by the vaso-dentine is larger than that formed by the cement (ib. c).

The third tooth (Plate XV. iii, Plate XVII. fig. 2, iii) is of nearly the same size and form as the second, but is somewhat narrower; the anterior and outer angle is less rounded off, and the external longitudinal depression is deeper: it is further removed from the second tooth than this is from the first.

The fourth molar (Plate XV. iv, Plate XVII. fig. 2, iv) is smaller than the two preceding, but of nearly equal length, viz. $8\frac{1}{2}$ inches, and is distinguished from the other teeth by being curved in only one direction, and that in a very slight degree, the concavity looking, as in the other teeth, outward: the central axis of the tooth, in reference to the anterior and posterior planes of the skull, is straight: the anterior and posterior layers of cement decrease in thickness as they approach the base of the tooth, so as to describe a slight curve, the convexity of which is turned, on both sides, towards the adjoining tooth. The fourth molar is tetragonal, and with more equal sides than the two preceding teeth; the outer and inner sides are concave, the anterior and posterior ones convex; the angles are rounded, but the anterior and inner angle is more produced than the rest. The grinding surface presents two equal transverse ridges, the contiguous sides of which are the longest.

The fifth molar (Plate XIV. fig. 1, Plate XV. v, Plate XVII. fig. 2, s) is 5 inches in length, 1 inch 2 lines in transverse, and $10\frac{1}{2}$ lines in antero-posterior diameter: its principal curvature presents its concavity forward, or toward that of the anterior tooth; the curve in the transverse axis of the skull is scarcely appreciable. The transverse section of this tooth is rhomboidal, with the angles rounded, and with the longest diameter intersecting the antero-internal and the postero-external angles. The dentinal axis is transversely quadrilateral, with the posterior angles entire, and the posterior surface concave: the layer of cement which covers this surface is the thickest, and its posterior surface is convex: the layers which cover the outer and inner sides of the tooth are, as in the rest, the thinnest; the anterior layer is less than one-third the thickness of the posterior layer. The anterior ridge of dentine is slightly prominent, and the posterior alone forms the summit of a transverse eminence with sloping sides, but these diverge at a more open angle than in the preceding teeth.

At the date of the publication of my 'Odontography,' no specimen of the lower jaw of the Megatherium had reached England, and certain detached teeth with slight differences from those known to belong to the upper jaw were conjecturally referred to the lower one[*]. The entire bone, with the dental series complete (Plate XVI.,

[*] Odontography, p. 341.

Plate XII. fig. 2, *i, ii, iii, iv*), shows that three of those teeth were rightly so referred; but that the small molar alluded to at p. 342, *op. cit.*, does not belong to the lower jaw, which has only four teeth in each ramus.

The first molar (Plate XVI. and XII. fig. 2, *i*) is 8 inches in length, with a pulp-cavity of 5 inches in depth; it presents a curve, with the convexity forwards, which is more marked than in any of the upper molars. The anterior surface is so much less convex transversely than in the first upper molar, that the transverse section presents a tetragonal rather than a semicylindrical figure; the anterior side, however, being only three-fourths the breadth of the posterior one, by which the first lower molar may be distinguished from all the tetragonal teeth of the upper jaw. Both the inner and outer sides are slightly concave transversely, the posterior side is moderately convex. The posterior ridge has a base twice as thick as the shorter anterior ridge. The greatest transverse breadth of the crown is 2 inches, the greatest fore and aft breadth is 1 inch 7 lines.

The second molar (Plates XVI. and XII. fig. 2, *ii*) is the largest, at least the broadest transversely, of those of the lower jaw. It is 9 inches in length, with a minor curvature, convex forward. The anterior side is the broadest, being more extended inward than the posterior side: its transverse diameter is 2 inches 3 lines, the fore and aft diameter of the crown is 1 inch 10 lines: the base of the hinder eminence in the latter diameter exceeds that of the front eminence chiefly by the greater extent of dentine exposed.

The third molar (Plates XVI. and XII. fig. 2, *iii*) is of the same length as the second, but has its two diameters more nearly equal, the transverse section being nearly square, the anterior division being rather the broadest transversely, and of equal thickness from before backwards. Both this and the preceding tooth are convex transversely before and behind, concave at the sides.

The last lower molar (Plates XVI. and XII. fig. 2, *iv*), with an equal antero-posterior diameter to the preceding, is shorter and narrower transversely, especially in regard to its posterior division, which is more rounded, or convex transversely, behind, than in any of the antecedent teeth. The hinder slope of the hinder ridge is more nearly horizontal, and those towards the middle of the tooth are less deep: the modification of the grinding surface of this tooth relating to the flatter surface of the fifth molar above, and its greater antero-posterior extent as compared with its breadth compensating for the absence of a fifth molar in the lower jaw. The grinding surface of the four lower molars equals that included between the anterior ridge of the first molar and the posterior ridge of the last molar in the upper jaw.

Each molar has its base undivided, but excavated by a deep conical pulp-cavity (Plate XVII. fig. 2, *p. p*), from the apex of which cavity a fissure is continued to the middle of the grinding surface of the tooth, where it is more conspicuous in the upper (Plate XV.) than in the lower molars. Plate XVII. fig. 2, exhibits a longitudinal section of the five molars of the upper jaw, *in situ*. The central axis of vaso-dentine (*v*) is surrounded by a thin layer of true or hard dentine (*d*), and this is coated by cement (c, c), which is of great thickness on the fore and hind surfaces, but is thin where it covers the outer and inner sides of the tooth.

As the outer layer of the vaso-dentine is first formed by the centripetal calcification of the pulp, the thin crust of that substance at the open base of the tooth includes a space equal to the vaso-dentine at the crown of the tooth: the contraction of the base of the tooth is due to the progressively-diminishing thickness of the cement as it approaches that part; the intervening vacancy (m, m) in the socket indicating the primitive thickness of the vascular capsule, by the ossification of which the cement is formed.

The vaso-dentine (Plate XVII. fig. 3, v) is traversed throughout by medullary canals, $\frac{1}{500}$th of an inch in diameter, which are continued from the pulp-cavity and proceed, at an angle of 50° to the plane of the hard dentine, parallel to each other, with a slightly undulating course, having regular interspaces, equal to one diameter and a half of their own area, generally anastomosing in pairs by a loop (ib. l, l), the convexity of which is turned toward the origin of the tubes of the hard dentine, forming a continuous reflected canal.

The loops are situated near, and for the most part close, to the hard dentine. In a few places one of the medullary canals may be observed to extend across the hard dentine, and to anastomose with a corresponding canal in the cement. The interspaces of the medullary canals of the vaso-dentine are principally occupied by dentinal tubes, which have an irregular course, form reticulate anastomoses, and terminate in very minute cells, at least one hundred times smaller than the calcigerous radiated cells of the cement.

The more regular and parallel tubes, which traverse the thin layer of unvascular dentine (ib. d), are given off from the convexity of the terminal loops of the medullary canals. The course of these tubes is more directly transverse to the axis of the tooth than is that of the medullary canals from which they are continued. They run parallel with each other, but with fine undulations throughout their course. They have a diameter of $\frac{1}{10,000}$th of an inch, and have interspaces of about twice that diameter. As the dentinal tubes approach the cement they divide and subdivide, and become more wavy and irregular; their terminal branches take on a bent direction and form anastomoses, dilate into small cells, and many are seen to become continuous with the radiating tubes of the cells of the contiguous cement.

The cement (ib. c), which enters so largely into the composition of the grinders of the Megatherium, is characterized in that extinct animal by the size, number, and regularity of the vascular or medullary canals (ib. m, m) which traverse it. They present the diameter of $\frac{1}{100}$th of an inch, and are separated by intervals equal to from four to six of their own diameters. Commencing at the outer surface of the cement, they traverse it in a direction slightly inclined from the transverse axis towards the crown of the tooth, running parallel with each other; they divide a few times dichotomously in their course, and finally anastomose in loops, the convexity of which is directed towards, and in most cases is in close contiguity with, the layer of hard dentine. Fine tubules are sent off, generally at right angles, from the medullary canals, which quickly divide and subdivide, form anastomosing reticulations, and communicate freely with the similar tubules that radiate from the lacunæ or calcigerous cells, ib. *r, r*. These cells are dispersed throughout the dentine, and present an oblong form, with the long axis transverse to that of the tooth, measuring $\frac{1}{7000}$th of an inch in diameter. The cavity of the cell, which is not quite occupied by their opake contents, is often very clearly demonstrated. The tubes, which radiate from the cells nearest the hard dentine, and from the terminal loops of the vascular canals, intercommunicate freely with the tubes of the hard dentine. The tooth of the Megatherium thus offers an unequivocal example of a course of nutriment from the dentine to the cement, and reciprocally.

In the structure which the fossil teeth of the Megatherium and its extinct congeners clearly demonstrate, we have striking evidence of the rich organization of those once-deemed extravascular parts, and that they were pervaded by vital activity. All the constituents of the blood freely circulated through the vascular dentine and the cement, and the vessels of each substance intercommunicated by a few canals continued across the hard or unvascular dentine.

With respect to those minuter tubes, the more important as being more immediately engaged in nutrition, which pervade every part of the tooth, characterizing by their difference of length and course the three constituent substances, they form one continuous and freely intercommunicating system of strengthening and reparative vessels, by which the plasma of the blood was distributed throughout the entire tooth for its nutrition and maintenance in a healthy state[*].

The grinding surface of the molars of the Megatherium differs, on account of the greater thickness of the cement on their anterior and posterior surfaces, from those of all the smaller Megatherioids, in presenting two transverse ridges; one of the sloping sides of each ridge being formed by the cement, the other by the vascular dentine; whilst the unvascular dentine, as the hardest constituent, forms the summit of the ridge, like the plate of enamel between the dentine and cement in the Elephant's grinder. The great length of the teeth and concomitant depth of the jaws, the close-set series of the teeth, and the narrow palate, are also strong features of resemblance between the Megatherium and Elephant in their dental and maxillary

---

[*] The first statement of the continuation of filamentary processes of the pulp into the tissue of the growing tooth was published in the 'Comptes Rendus de l'Académie des Sciences,' Paris, 1839, p. 787, and the earliest observation of their continuation into the dentinal tubuli was, I believe, recorded in the following passage:— "I had the tusk and pulp of the great Elephant at the Zoological Gardens longitudinally divided, soon after the death of that animal in the summer of 1847. Although the pulp could be easily detached from the inner surface of the pulp-cavity, it was not without a certain resistance; and when the edges of the co-adapted pulp and tooth were examined by a strong lens, the filamentary processes from the outer surface of the pulp could be seen stretching as they were withdrawn from the dentinal tubes before they broke."—Art. TEETH, Cyclopædia of Anatomy, vol. iv. p. 929.

organization. In both these gigantic phyllophagous quadrupeds provision has likewise been made for the maintenance of the grinding machinery in an effective state; but the fertility of the Creative resources is well displayed by the different modes in which this provision has been effected: in the Elephant, it is by the formation of new teeth to supply the place of the old when worn out; in the Megatherium, by the constant repair of the teeth in use, to the base of which new matter is added in proportion as the old is worn away from the crown. Thus the extinct Megatherium had both the same structure and mode of growth and renovation of the molar teeth, as are manifested in the present day by the diminutive Sloths.

§ 6. *Comparison of the Skull and Dentition.*

The important affinity indicated by the dentition is confirmed by the characters of the skull. In no other edentate family, save the *Bradypodidæ*, is the cheek-bone so nearly developed to the megatherioid proportions of that bone; in no other does it ascend above the zygoma into the temporal fossa or descend below the level of the molar teeth. The large and complex malar bone is also associated, in the Sloths, with a terminal position of the great anterior and posterior orifices of the cranium, with terminal occipital condyles, and in the Ai (*Bradypus tridactylus*) with a sloping occipital region. The cranial division of the skull is relatively as great in the Sloths as in the Megatherium; and the actual capacity of the cerebral cavity is masked by a similar expansion of the air-cells, which almost everywhere surround that cavity, and raise the outer plate of its bony parietes above the inner one. The occiput presents the same expanded proportions, the same broad depressed basilar plate, and the anterior condyloid foramina are of large relative size. The tympanic is small; it nearly completes a circular frame for the ear-drum, to which function it is limited, and it long remains a separate bone. The detached and, in the skull described, lost tympanics of the Megatherium have been evidently restricted to the same office. The temporal fossa, in the Sloths, is long and large, and communicates freely with the orbit, the outer boundary of which, however, is not completed in any living species of Sloth. The nasals become confluent in old Sloths, and develope turbinal laminæ from their under surface. The premaxillaries are edentulous and without any ascending process. The rami of the lower jaw expand and branch out behind into a coronoid, a condyloid, and a long and deep angular process, and they are anchylosed anteriorly at a broad sloping symphysis. Only in the genus *Bradypus*, amongst known existing quadrupeds, do the alveoli of both jaws correspond in number, simplicity, relative depth and position with those of the Megatherium. The still more important agreement between these existing and extinct *Bruta*, in the peculiar structure of the teeth, yields the crowning proof that it is to the diminutive arboreal Sloths that the gigantic Megatherium and its less bulky though larger extinct congeners have the closest natural affinity.

The chief differences observable in the cranial anatomy of the Sloths, as compared with that of the Megatherium, are the greater relative depth and breadth, and the more convex outline of the coronal aspect of the skull; but this difference would be, doubtless, much less marked in the immature than in the adult Megatherium. The zygomatic process, in the Sloths, is relatively shorter, and does not attain the malar bone; this, therefore, has not the middle process for supporting the zygoma, and is two-pronged, instead of being, as in the Megatherium, four-pronged. The chief characters by which the Megatherium deviates, in its cranial structure, from the bradypodal and approaches to the myrmecophagal type, are the elongation of the slender edentulous fore-part of the upper jaw, and of the corresponding grooved slender edentulous part of the lower one: but the prolongation of the upper jaw is due to relatively longer premaxillaries than are developed in any of the true Edentata. The zygomatic arches, moreover, are more defective in the Anteaters and Pangolins than in the Sloths; the malar part especially being minute or obsolete. Only in the Orycterope and Armadillos, amongst the existing *Bruta*, is the zygomatic arch complete, but it is simple, without ascending or descending processes. The great Glyptodon, indeed, exemplifies that tendency to community of characters so often presented by extinct species, in an inferior prolongation of the malar bone analogous to that of the Sloths and Megatherium. With other existing mammals than those of the Edentate order, it would be lost time to pursue the present comparison with a view to the elucidation of the affinities of the Megathere. It needs only to place the skull of this animal by the side of those of the Elephant, Rhinoceros, Sivatherium, Ox, Elk, Horse, Dugong, or other vegetable-feeding mammal of corresponding or approximate size, to be struck with the peculiarities of the fossil, and **to be convinced that the habits and mode of** feeding of the Megatherium had been such as are no longer manifested by the larger Herbivora of the present day.

It remains then to inquire whether, among the extinct forms of the mammalian class to which was assigned the office of restraining the too luxuriant vegetation of a former world, there be any that, from their cranial or dental characters, may be concluded to have resembled the Megatherium in the mode of performing that task.

The skull of the Mylodon, while it presents all the essential resemblances to that of the Megatherium which have been pointed out in the skull of the Sloth, as, *e.g.* the long cranium, the terminal position of the occipital condyles, and of the occipital and nasal apertures, and the large and complicated malar bones, approximates still more closely to the Megatherium in the junction of the malar with the zygomatic process of the temporal, and in the relative depression and flatness of the elongated cranium. But in thus receding from the existing Sloths, the Mylodon does not approach any other existing genus, but only another member of its own peculiar extinct family.

The most marked differences in the skulls of the Mylodon and Megatherium depend on the minor length of the teeth and consequent depth of their sockets in the smaller species, which require a less vertical extent of the maxillary bone in the molar region and of the corresponding part of the lower jaw, the lower border of which is consequently nearly straight in the Mylodon, as it is in the existing Sloths. So great a proportional extent of the **descending process of the malar bone is consequently not required**, and this **process is more oblique** in direction, and is **relatively broader and thinner** than in the Megatherium. In these differences also the Mylodon shows its closer resemblance to the Sloths. The basioccipital is relatively broader, and the occipital condyles are wider apart in the Mylodon; the occipital plane is more inclined and the zygomatic process proportionally weaker than in the Megatherium. The minor depth of the lower jaw and the lighter grinding instruments call for **a less extensive** origin of the temporal muscles, and accordingly the superior boundaries **of their fossæ** are separated in the adult Mylodon by a wide and smooth parietal tract*, as in the Sloths. The postfrontal process is rudimentary in the **Mylodon, and the postorbital process of the malar bone is obsolete; the orbit is consequently without any bony boundary behind; the malar, accordingly, has but three processes, and is thus** less complicated than in the Megatherium. The interorbital part of the skull is relatively narrower, the maxillary part relatively broader, than in the Megathere. No trace of premaxillaries was **present in the skull of the** *Mylodon robustus* described by me, and the broad truncated **symphysis of the lower jaw indicates that they must have been very small if they existed:** the peculiar length of the premaxillaries in the Megatherium, and the corresponding prolongation **of the** long and narrow symphysis mandibulæ, offer the most **conspicuous differences in the** conformation of the skull, and proportionally **remove that genus from the existing** Sloths. The bony palate **is absolutely broader in the much smaller skull of the** Mylodon than in the Megatherium, and **it gradually** contracts from the first to the fifth molar: **it is,** *e. g.* **5 inches in breadth between the** first molars in the *Mylodon robustus*, **and only 2 inches in breadth between the** same teeth in the Megatherium.

The opportunity of a comparison of the skull of the Megalonyx with that of the Megatherium is yet a desideratum: it would probably demonstrate some intermediate modifications between the latter and the Mylodon.

The extinct megatherioid animal of which, after the Mylodon and the Megatherium, the most complete cranium has **hitherto** been obtained, is the Scelidotherium†.

In many respects the skull of this animal resembles that of the Megatherium more than does that of the Mylodon. The plane of the occiput is rather less inclined from below forward than in the Megatherium, but more resembles that part than in the Mylodon: the upper boundaries of the temporal fossæ more nearly approximate **than in the Mylodon**: the bony palate is narrower, and its sides more parallel than in the **Mylodon;** but instead of being concave transversely, as in the Megatherium, it is convex: the alveoli are nearer together than in the Mylodon, and the first is not separated by a wider diastema than the rest. The symphysis of the lower jaw is much

* Memoir on the Mylodon, 4to, 1842, pl. 3.
† Fossil Mammalia of the 'Voyage of the Beagle,' p. 73, pls. 20, 21, & 22.

prolonged, but is less deeply channelled above than in the Megatherium, and is not so distinctly defined by the abrupt increase of depth of the ramus behind it which characterizes the Megatherium; the molar part of the mandible makes, however, a greater convexity below than in the Mylodon.

With these marks of approximation to the Megatherium there are, however, the same differences as in the Mylodon in regard to the widely open orbit, the more simple, trifurcate, malar bone, the minor depth of the alveolar portions of the jaws, and the straighter outline of the lower border of the mandible. In both the Mylodon and Scelidothere the coronoid process is relatively shorter than in the Megathere, and the foramen near the fore-part of the base of that process is outside and below that base, not on the inside of it as in the Megatherium. The mastoid process is relatively shorter and the stylohyal pit is shallower; the lacrymal bone is more distinct, and the foramen is larger in the Scelidothere than in the Mylodon or Megatherium. In all the essential characters of the lower jaw, as in the number, structure and kind of teeth, the extinct megatherioid quadrupeds more closely resemble each other, and the existing Sloths, than any other known existing or extinct animals.

The number of the teeth, their length, equable breadth and thickness, and absence of fangs, their deeply excavated base, and the unlimited growth resulting from the persistent matrix, together with their composition of cement, dentine and vaso-dentine, without any true enamel, are characters common to the Megatherioids and Sloths. The form of the teeth differs in, and characterizes, each genus. It would seem that the Megalonyx, in the elliptical or subcylindrical shape of such of its teeth as are known, more closely resembled the existing Sloths than do the other Megatherioids. The Unau (*Cholœpus didactylus*) resembles the Mylodon in the distance between the first and the second molars of the upper jaw; but the advanced molar assumes, in that existing Sloth, the form and proportions as well as the position of a canine, and the corresponding tooth of the lower jaw is similarly developed and separated from the other three teeth by a nearly equal interval. In the Ai (*Bradypus tridactylus*) the first molar in both jaws presents nearly the same proportionate size to the rest, as in the Megatherium, and is not separated from them by a wider interval than the rest, as it is in the Unau and Mylodon. In both species of existing Sloth, the last molar of the lower jaw is as simple in form as in the Megatherium: in the Mylodon and Scelidotherium it is larger and more complex in shape, the grinding surface being divided into two lobes by two oblique channels, which traverse longitudinally one the outer the other the inner side of the tooth: these grooves are more shallow in the Scelidothere than in the Mylodon, and the lobes of the tooth are more equal and more compressed. The grinding surface itself, in all the molars of both Mylodon and Scelidothere, resembles that in the Sloths; the two transverse ridges are developed only in the teeth of the Megatherium, which are longer in proportion to their thickness than in the Mylodon, Megalonyx, or Scelidothere. These modifications, with the narrow palate, the close-set series of teeth, their great length, and the concomitant depth of the jaws, are features of resemblance to the maxillary and dental characters of the Elephant; but the fundamental structure of the teeth, not only of the Megatherium, but of all its extinct congeners, is manifested in the present day exclusively by the restricted and diminutive family of the *Bradypodidæ*. I conclude, therefore, the present section of this memoir by repeating the remark which I was led to make in a former memoir*, relative to the existing Sloths:—"These Mammals present to the zoologist, conversant only with living species, a singular exception in their dental characters to the rest of their class; but there has been a time when this peculiar dentition was manifested under as various modifications as may now be traced in some of the more common dental types in existing orders of Mammalia."

Comparative Table of Dimensions of the Skull of the MEGATHERE, MYLODON, and SCELIDOTHERE.

	Megatherium Americanum.			Mylodon robustus.			Scelidotherium leptocephalum.		
	ft.	in.	lin.	ft.	in.	lin.	ft.	in.	lin.
CRANIUM.									
Length from the occipital condyles to the fore-end of the upper jaw	2	7	0	1	6	6	1	8	4
Length from the occipital condyles to the fore-part of the malar bone	1	8	0	1	2	0	0	11	6
Length from the fore-part of the malar bone to the fore-end of the upper jaw	0	10	6	0	5	0	0	8	8
Breadth across the widest part of the zygomatic arches				0	10	9	0	7	0
Least breadth at the interspace of those arches	0	6	0	0	5	4	0	3	7
Breadth of the fore-part of the nasal bones	0	4	6	0	3	3	0	2	8
MANDIBLE.									
Length	2	3	0	1	3	6	1	6	6
Breadth between the hinder ends (angles) of the rami	0	7	0	0	6	3	0	4	7
Breadth between the condyles	0	4	0	0	4	2	0	1	8
Breadth between the posterior sockets of the teeth	0	2	3	0	2	10	0	1	6
Breadth between the anterior sockets of the teeth	0	1	9	0	4	0	0	1	8
Breadth across the fore-part of the symphysis	0	4	5	0	5	4	0	2	5
Depth of ascending ramus from the upper part of the condyle	1	5	0	0	5	7	0	5	8
Depth of ascending ramus at the fore-part of the base of the coronoid process	0	9	0	0	3	7	0	3	7
Depth of horizontal ramus at the fore-part of the first socket	0	8	0	0	3	0	0	2	7
Length of the symphysis following the outer curve	1	1	0	0	4	3	0	7	3
Fore and aft extent of base of coronoid process	0	9	0	0	3	8	0	3	9
From the back part of the condyle to the end of the angular process	0	6	0	0	3	2	0	3	6
From the end of the angular process to the last socket	0	9	7	0	6	6	0	6	5
From the first socket to the anterior margin of the jaw	0	9	0	0	3	6	0	7	10
Extent of the alveolar series	0	9	0	0	5	4	0	4	4
Breadth of the condyle	0	3	0	0	2	6	0	1	10

§ 7. *Of the Bones of the Anterior Extremity.*

The bones of the limbs of the Megatherium are not less fraught with interest to the comparative anatomist and physiologist than are those of the trunk and head, by reason of their peculiar proportions and configurations, and, more especially, as the unguiculate type on which they are constructed is exemplified in a quadruped of such enormous bulk. The anterior extremities exceed the posterior ones in

* On the *Mylodon robustus*, 4to, p. 43.

length: in their bony structure they include a complete clavicle with the scapula, a humerus, an antibrachium consisting of fully developed and reciprocally rotating radius and ulna, the carpus, metacarpus and four digits; and manifest, in short, all the main perfections of brachial structure, save the opposable thumb, observable in the Mammalian class. These perfections, moreover, are associated with proportions and processes indicative of enormous strength, and bespeak a limb fitted not only to take its full share in the support of the body, but to be employed in operations in which unusual resistance has manifestly to be overcome. In no respect, perhaps, does the Megatherium more strikingly differ in its osseous structure from the existing quadrupeds of corresponding bulk, than in the bony fulcra of the anterior extremity.

The scapula (Plate I. м, and Plate XVIII. figs. 1 & 2) presents a vast expanse of bone, with a double spinous process; the normal one expanding into a large acromion which is continued into and is confluent with the coracoid process. The scapula presents an inequilateral triangular form, of which the acromion is the apex; the main body or plate of the bone approaches to the form of the geometrical trapezium, except that the extent of the posterior border or base is such as to destroy the parallelism of the upper and lower borders. The upper border (b, c) is the shortest; but, in one specimen, owing to the greater development of the basal border it appears to begin, as in Plate I., at the part of the base marked a in Plate XVIII., and to form a low angle, as if continued about one-fourth of the distance from the base parallel with the lower border, whilst the rest of the costa inclines downwards or converges towards the neck of the scapula, with a slight concave outline. The upper border increases in thickness as it passes into the origin of the coracoid, c. The base of the scapula, from the point a, is straight as far as the origin of the spine; it then bends, with a convex curve, and increases in thickness to the inferior angle of the scapula (d) close to which commences the second or lower spine. The inferior costa of the scapula (d) extends forwards straight and parallel with the lower spine, for some way, and is so continued into that spine that the fore-part of this might be regarded as the inferior and anterior angle of the scapula bent outwards; but it curves down to what may be more correctly considered as that angle at e, fig. 1, Plate XVIII.

The normal spine of the scapula, notwithstanding its being the superior of the two, commences at only one-fourth of the entire base of the bone from the inferior angle; it is thence continued forwards, gradually rising or gaining depth, parallel with the inferior costa, and thus marks out a very large proportional extent of the outer surface of the bone for the supra-spinal fossa, Plate I. м. As the spine advances it increases in breadth as well as depth, until it springs clear of the main body of the bone as the acromial process, k. This large, thick and rough process arches forward and upward over the glenoid articular cavity (g), and meets and coalesces with the coracoid (c), spanning the anterior outlet (f) of the supra-spinal fossa by a strong and broad bridge of bone. Through the reciprocal modification of the coracoid, the passage for certain vessels and nerves, which is usually in Mammals a mere notch of the upper border of the scapula, is converted into a complete foramen, h.

The second spine, which answers to the ridge defining the upper part of the fossa for the 'teres major' muscle in Man, runs parallel with the upper spine, and then bends down to terminate in the low angle (Plate XVIII. fig. 1, *e*), projecting from the inferior costa of the scapula, about one-third of the length of that border from the glenoid cavity (*g*). The shape of that cavity is almost an ellipse (Plate XVIII. fig. 2, *g*), the lower end being scarcely more contracted than the upper one: both these parts of the periphery are rather more produced than the lateral borders. The entire margin of this moderately deep and well-defined articular cavity is thin, but convex. The inner surface of the scapula (Plate XVIII. fig. 1) is divided into many shallow depressions by intermuscular ridges (*i, i*), having a general tendency to converge from the basal border, where most begin, towards the centre of the bone, though none extend so far. The inferior costa (*d*) is continued more directly towards the inner side of the glenoid cavity by a smooth, convex rising of the inner surface, which grows broader as it gradually subsides: the fore-part of the lower costa (fig. 1, *e*) is rather a continuation of the inferior spine; between which and the portion of the lower costa (*d*), is the wide channel, attesting the size and force of the homologue of the *teres major*. The vigour of the enormous *subscapularis* muscle is manifested by the intermuscular ridges (*i, i*).

Compared with the scapula of the Mylodon, that of the Megatherium differs in the lower position of the spine, and the consequent greater expanse of the supraspinal fossa: the base is straighter and relatively longer, the inferior costa relatively shorter. The ridge extending from the inferior and posterior angle forwards upon the under surface of the scapula, is relatively stronger in the Megatherium, and gives to the homologue of the inferior costa in the Sloth's blade-bone the character of a second spine. As such 'second spine,' with a prolongation of the scapula below, characterizes that bone in the Anteaters, this indication of affinity to *Myrmecophaga*, or rather manifestation of the more general characters of the order *Bruta*, by the Megatherium, is not without interest. The confluence of the acromion with the coracoid is peculiar to the Sloths amongst existing Mammals.

Clavicle.—Before the discovery of the Megatherium, Man was usually cited in Manuals of Comparative Anatomy as the largest animal possessing a 'collar-bone': he is, in fact, the largest existing animal so endowed. But whilst the length of the human clavicle averages but 6 inches, that of the Megatherium is 15 inches. In its general shape and sigmoid curve, this bone (Plate XIX. fig. 1) singularly resembles the human clavicle, but is thicker in proportion to its length. No single bone would have better excused the common conclusion of the mediæval anatomists as to the nature of large fossil bones, viz. that they were those of human giants, than the collar-bone of the Megatherium.

The sternal end is expanded, obliquely truncate, with a rough, irregular, undulating, but mainly convex articular surface, much fitted for strong ligamentous attachment to the manubrium sterni, and with a narrow, rather flattened facet above this surface, where it abuts against the same part in its fellow, as shown at Plate XXVII. *u*.

The shaft of the bone is most contracted at its sternal third part; thence it gradually expands to the acromial end; the anterior surface is moderately smooth and convex: the posterior surface (Plate XIX. fig. 1) is rough and more flattened, and is traversed obliquely by a broad ridge, which terminates outwardly, or developes the rugged upper border of the acromial half of the bone. This expanded end bends downward, as the sternal end bends upward. There is a large tuberosity on the outer or fore-part of the acromial expansion, which terminates in an oblong convexity, adapted to the concavity beneath the expanded end of the acromion. The strong aponeurotic character of the periosteum of the clavicle is well shown by the linear decussating ridges on most parts of the surface of the shaft. The bone is solid.

The closer affinity of the *Bradypus didactylus*, as compared with the *Brad. tridactylus*, to the Megatherium, is illustrated by the complete clavicles which attach the scapulæ to the sternum; but they are straighter, relatively more slender, and more suddenly expanded at the sternal end than in the Megatherium. One species or variety of Three-toed Sloth has only a small styliform clavicular bone, appended to the coracoid process. The clavicle in *Orycteropus* and *Myrmecophaga* is complete, but has a single curvature. The clavicle of the Mylodon is intermediate, in its form and proportions, between that of the Megatherium and that of the Two-toed Sloth.

The supposed peculiarity in the articulation of the clavicle of the Megatherium with the first rib instead of the sternum, which CUVIER inferred from the figures and descriptions of the fossil animal which had been published in his time*, is shown, by the more perfect specimens since received, not to exist in nature; and the suspicion expressed by the great anatomist, viz. that it might be due to some misarticulation in the Madrid skeleton, is thus proved to be well-founded. The true connexions of the sternal ends of the clavicles with each other and with the manubrium sterni, are well shown in the view of the skeleton in the British Museum, given in Plate XXVII.

Humerus.—The humerus (Plate XIX. figs. 2—5) is remarkable for the vast expanse of the lower fourth part of the bone; but this is limited to the transverse direction; so that, viewed sideways (as in Plate I.), the humerus of the Megatherium appears to be a comparatively weak and slender bone: the whole shaft, however, gives indications of the force of the thick muscular masses which surrounded and operated on it.

The head of the bone (Plate XIX. fig. 4) presents a smooth convexity of an elliptical form, corresponding with that of the glenoid cavity of the scapula; the long axis of the ellipse is from before backward. The head rises clear above the outer and inner tuberosities, neither of which are so developed as to interfere, as in ungulate quadrupeds, with the free rotation of the bone.

The rugged surface of the inner tuberosity (*ib.* figs. 2 and 3, *a*) slopes downward and inward (ulnad) from the peripheral groove of the head, marking the attachment of the

---

* "D'après les figures et les descriptions, il paraîtrait que cette clavicule s'articulerait, non pas avec le sternum comme à l'ordinaire, mais avec le bas de la première côte qui est recourbée, et présente une concavité pour la recevoir. Ce serait une singularité dont je ne connais pas d'exemple."—Ossemens Fossiles, Ed. 1835, 8vo, tom. viii. p. 349.

joint-capsule. The pectoral ridge is continued from the lower, slightly outstanding, part of this tuberosity. The outer tuberosity (ib. *b*) projects from a lower level, is larger and more prominent than the inner one; and is divided into two subequal rough facets. The outer deltoid ridge begins from the outer side of the tuberosity; the inner deltoid ridge (ib. *d*) some inches lower down, and nearer the inner side of the shaft: these ridges converge, strengthen as they descend, and coalesce at the beginning of the lower third of the shaft, defining a long and narrow angular tract for the implantation of the powerful deltoid muscle. A tuberosity (ib. *c*) is developed from the outer (radial) side of the humerus, one-third down the bone, from which a strong ridge descends along the same side of the middle third: this ridge is defined below, and divided from the supinator ridge, by a deep and smooth oblique channel (ib. *e*) for the passage of vessels and nerves from the back to the fore-part of the bone. The supinator ridge (ib. *f*) is the most prominent feature of the lower third of the humerus; it presents a long, triangular, rough outer facet, widening as it descends, and with a secondary ridge from its middle part. The longitudinal contour of this facet forms an obtuse angle with the lower half of the ridge, extending to the outer condyle of the humerus; the outer facet of this half is rough, triangular, with the base upward. The pectoral ridge (*ib.* fig. 3, *i*) terminates in a low tuberosity (ib. *h*) on the inner side of the middle of the shaft; whence a second ridge is continued upward upon the back of the shaft. This surface is flatter than the fore-part, especially at its lower expanded third; at the bottom of which, midway between the outer and inner supracondyloid productions, and just above the lower articular surface, is a small but well-defined olecranal depression. The inner supracondyloid angular production (ib. *k*) has the flat rough facet only upon its lower half. Owing to the production of these ridges, the articular condyles themselves (ib. *g*, *l*) appear to occupy but a small part of the distal end of the bone; for the extreme breadth of this end being 13 inches, that of the articular surface is but 7½ inches. It consists of two convexities, side by side, divided by a narrow and deep channel, continued from the front non-articular surface half-way towards the back part of the bone. Both convexities have a full elliptic periphery; the outer one (*ib.* figs. 2, 3, 5, *g*) with the long axis from before backward, the inner one (ib. *l*) with the long axis from side to side: the outer condyle is the larger and more prominent of the two; it forms more than a hemisphere, the antero-posterior contour describing full three-fourths of a circle. The extent of flexion and extension of the fore-arm on the arm is thus shown to be considerable. The articular surface continued from one condyle to the other is concave transversely.

The centre of the shaft of the humerus is occupied throughout by a coarse cancellous structure. A very small medullary artery penetrates the back part of the bone, below the pectoral tuberosity, the canal extending obliquely downward and outward.

The *Myrmecophaga didactyla*, amongst existing *Bruta*, most resembles the Megatherium in the development of the supinator or outer supracondyloid ridge.

In its general proportions, the humerus of the Megatherium resembles that of the Megalonyx; and is more slender, in proportion to its length, than in the Mylodon or

Scelidothere. The articular head forms a larger proportion of a sphere, and projects more freely beyond the tuberosities: these are relatively smaller, and are more equal than in the Mylodon or Scelidothere: the external tuberosity, in particular, is more developed in these smaller Megatherioids. In them also the external ridge is continued from above the 'musculo-spiral' groove inwards, along the front of the humerus to the apex of the deltoidal tract, forming its outer boundary: in the Megatherium, a smooth concave surface divides the outer ridge from the deltoidal elevation, which is absolutely narrower. The vertical outline of the back part of the shaft of the humerus in the Megatherium is almost straight, being but a little bent forwards at its lower third, as it is likewise in the Megalonyx; and, in both, the olecranal depression is well defined: in the Mylodon and Scelidotherium, the same outline of the humerus is slightly concave; the lower third of the bone being, as it were, a little bent back below the deltoidal platform; and the olecranal fossa is not defined. The inner supracondyloid plate is produced at its upper part into a strong tuberosity, in both the Mylodon and Scelidotherium, but not in the Megatherium.

In the existing Sloths, the humerus at this part, viz. above the inner condyle, is perforated in one genus (*Cholœpus*) and not in another (*Acheus*, F. CUVIER): and the same difference occurs in the great extinct Sloths. In the Megatherium the humerus is imperforate, as it is in the Mylodon: in the Megalonyx and Scelidotherium it is perforated above the inner condyle. Yet the Megalonyx most resembles the Megatherium, not only in the general proportions of the humerus, but in the configuration of its two articular ends. The inner or ulnar condyle, *e. g.*, is convex in every direction in the Megalonyx as in the Megatherium: in the Mylodon and Scelidotherium it is convex only from before backwards, but is concave from side to side. In the more robust proportions, in the shape of the articulations and the development of the processes of the humerus, the Mylodon and Scelidotherium as closely resemble each other, as the Megatherium resembles the Megalonyx in the same characters; yet the humerus of the Scelidotherium has the inner perforation, and that of the Mylodon a groove merely, for the brachial artery and nerve. This variety, and the corresponding one above noticed between the Megatherium and Megalonyx, show the inapplicability of the final cause commonly assigned to the ento-condyloid hole, the existence or otherwise of which depends merely on the ossification or non-ossification of the aponeurosis extending from the shaft of the humerus to the ento-condyloid process.

Ulna.—The bones of the fore-arm in the Megatherium, as in the Megalonyx, exemplify by their greater length, as compared with those of the same segment in the hind limb, the Bradypodal affinities of these huge extinct quadrupeds.

The ulna in the Megatherium (Plate XX. figs. 1, 2, 3) is, however, peculiar for the vast expanse of its proximal end, in connexion with its long and slender shaft. The olecranon (ib. *a*) is twice as broad as it is long; its inner border, springing from the notch which penetrates the inner and back part of the humeral articular fossa, extends obliquely **upward** and inward for 3½ inches, the ridge or edge **of the** plate being about an inch in thickness: the plate from this ridge sweeps round the back part of the bone, subsiding and increasing in breadth as it approaches the radial side, where it terminates in a tuberosity, divided by a groove from the radial articular cavity (ib. *b*), and prolonged downward into the ridge bounding the radial side of the upper half of the **ulna**. The olecranon projects rather backward than upward, and is strengthened and supported at its highest part by a strong angular buttress, which gradually subsides upon the back of the ulna.

The great 'sigmoid' articular surface is divided into two facets by a median portion, which is produced forward in the longitudinal or vertical direction, and is convex from side to side; the divisions so defined are **concave**. The inner one (ib. *c*), for the inner **humeral condyle, describes** a semicircle from behind forwards; it has little more than **half that** extent from side to side, and is encroached upon by a narrow, rather deep, rough channel, continued from the inner origin of the olecranon, expanding, to near the middle of the cavity. The outer division of the sigmoid fossa (ib. *b*) has reverse proportions, the transverse being nearly double the extent of the longitudinal diameter, **and it** is less concave than the inner division; about half an inch of its lower border bends a little back for the articular margin of the head of the radius, just above the rough fossa, for the reception of the non-articular part of the same head; the rest of the outer division of the sigmoid cavity receives the back part of the outer condyle of the humerus. Below the outer division the **radial fossa** (ib. *d*) **presents a triangular form**, bounded by a rough ridge externally and by a **tuberosity internally**. The ridge is continued upon the radial side of the shaft **of the ulna to within a fourth part of its distal** end; the surface of this side of the bone is irregularly sculptured, **indicating the strong** ligamentous connexion between the ulna and radius at this part. The **back and inner** sides of the shaft of the ulna are comparatively smooth.

A vascular canal, somewhat larger than the rest, is seen near the fore-part of the outer surface, entering **the bone** obliquely upward. A rising of the surface, with a linear series of three or four rough tubercles, marks the lower fourth of the inner side of the bone; a short wide longitudinal channel marks the back surface of the distal end (fig. 3), which is slightly expanded and convex, and so impressed as to indicate its ligamentous junction with the carpus.

The ulna of the Megatherium differs from that of the Mylodon, not only in its longer and more slender proportions, but also in the absolutely as well as relatively minor **height or** length of the olecranon; in the much less relative vertical or longitudinal extent of the outer division of the 'sigmoid' cavity; and in the 'haversian' fossa on the inner division. It differs in the much narrower channel dividing the articular cavity from that part of the base of the olecranon which is continued into the posterior ridge or border of the shaft; it differs, also, in the convexity of the distal end and the absence of the articular facet, which is distinctly present in that part of the ulna in the Mylodon and Scelidotherium.

Radius.—The radius (Plate XX. figs. **4, 5,** 6), like the ulna, of the Megatherium resembles in its longer and more slender proportions that bone in the Megalonyx, and differs from the proportionally thicker and shorter radius of both the Mylodon and Scelidotherium. The proximal end is circular, and is occupied by a smooth, moderately shallow, articular cavity (ib. fig. 5), with a well-defined border, over which the articular surface extends, on the ulnar side of the head, for about half an inch down; which tract is adapted to the lower portion of the outer division of the sigmoid cavity of the ulna. The articular modification of the head of the radius is as completely adapted for the superadded rotatory movements of the antibrachial bones, as in the human subject, to the head of the radius of which the resemblance of that of the Megatherium is strikingly close.

The shaft of the radius gradually narrows, in the antero-posterior diameter, along the upper fourth part, but maintains the same diameter as the head, transversely. Three inches below the head, on the inner and fore-part of the shaft, is the tuberosity (ib. *a*) for the tendon of the biceps, which measures 3 inches in long diameter and $1\frac{2}{3}$ inch across. Here the bone bends a little outward (radiad); and the ridge bounding that side is developed into what may be termed a process (ib. *b*), with a low angle, whence the ridge is continued straight down the lower half of the shaft to near the tuberosity above the styloid process (ib. *c*), where it curves outwardly to terminate in that tuberosity. **The fore-part of the** shaft is moderately smooth and convex across; it describes, lengthwise, a slight concavity; on the inner side of the bone, a broad and very rugged tract begins, about an inch and a half below the bicipital tuberosity, and extends along the middle third of the shaft; a less rough tract is continued thence, gradually expanding to the cavity for the lower end of the ulna. The outer side of the shaft of the radius is smooth, convex across, and with a slight convexity in its longitudinal contour: from the external process downward the radius maintains an equal breadth near the lower end, and **there it** expands in all directions to form the large articular cavity (ib. *d* and fig. 6) for the major part of the carpus. Above this cavity, on the ulnar side, is the rough and shallow depression for the ligamentous junction of the corresponding end of the ulna; on the front side a broad low ridge extends obliquely from the suprastyloid tuberosity to the border of the cavity; on the back part three oblong, short, thick ridges or tubercles divide the surface into four channels, for tendons; three of these are longitudinal and parallel, progressively increasing in width from the innermost (or one next the ulna) to the third; the outermost passes obliquely between the suprastyloid tubercle and the styloid process. This process is short and thick, rounded at the end; flattened in front, with the smooth articular surface continued for a few lines upon the lower border of this surface, from the general articular cavity which is extended over the lower end of the radius including the styloid process. This cavity presents a triangular form with the angles rounded off; the base next the ulna is short and oblique; the anterior border or side is the longest, the opposite border is the thickest, and is notched near the styloid process: the cavity is moderately concave, and articulates with the scaphoid and lunar bones of the carpus.

In the Mylodon, the radius is not only thicker in proportion to its length, but is more extensively and deeply impressed by the muscles of the fore-arm, especially on the back part of the bone. The tuberosity for the insertion of the biceps is further from the proximal joint, and augments the power of the muscle in the same degree: the proximal articular cavity is of an oval form.

Carpus.—The carpus (Plate XXI. s....u) consists of seven bones, four in the proximal and three in the distal row.

Scapho-trapezium.—The first of the proximal row (ib. s) includes the bone (ib. t) answering to the first of the distal row in Man and most Mammals, and is consequently a 'scapho-trapezium' (s, t)*, as it is also in the Sloths, the Mylodon and Scelidotherium. It is of an irregular triangular shape, with its base applied to the 'lunare' (ib. l), and with the apex somewhat twisted. It presents a broad convex articular surface (ib. s) for the outer half of the concavity of the radius; and this surface is continuous with a crescentic one of about one inch in breadth, which covers the proximal part of the side of the bone next the os lunare. The palmar or anterior horn of the crescent is continuous with an oval flat articular surface joining the os magnum (ib. g): the opposite or dorsal horn is separated by a rough tract from a convex subquadrate surface which also articulates with the os magnum. On the outer side of this surface, and, like it, on the fore-part of the bone, is a surface concave in one direction and convex in the opposite direction, for articulation with the small trapezoides (ib. z). External to this, the fore-part of the produced and twisted apex of the bone, which represents the trapezium, articulates with the stunted metacarpal of the pollex (ib. m 1), chiefly by ligament, but also by a small elliptical flat surface which seems to have been covered with articular cartilage. Between the two facets for the os magnum the bone is deeply excavated and has been perforated by blood-vessels.

Lunare.—The 'os lunare' (Plate XXI. l) offers, as in Man, some rude resemblance to a crescent; its proximal surface, very convex from before backward and rather convex from side to side, is wholly covered by the smooth articular surface which plays upon the ulnar half of the terminal cavity of the radius; and this surface is continued upon the radial side of the bone to form there the crescentic tract, adapted to the similarly-shaped tract on the scaphoid. The inner horn of this tract is continuous with the surface, convex at the fore-part, then deeply concave from before backwards, for the os magnum (ib. g): this articular surface is continuous with a similarly deeply excavated and irregular one on the ulnar half of the fore-part of the bone, which is subdivided into three facets, the middle one for the os cuneiforme, the two smaller ones for two parts of the os unciforme.

Cuneiforme.—The 'os cuneiforme' (ib. c), which is the smallest of the three proximal carpals, presents at its radial side a triangular convex surface for articulation with the

---

* This is the bone called 'cuneiforme' in Cuvier's chapter on the Megatherium, and which is marked *r* in the copy of Dr. Pander's figure of the Madrid skeleton introduced into plate 217, fig. 3, of the edition of the 'Ossemens Fossiles,' 8vo, 1835, here cited.

lunare; and at its fore-part an irregular concavo-convex oblong surface for the unciforme (ib. *u*). Its proximal surface is tuberous and rough for ligamentous attachment to the ulna, except where the smooth articular surface for the os pisiforme is situated: this surface is **in great part** convex.

In the Mylodon the os cuneiforme is the **largest of the carpal** bones.

Pisiforme.—The os pisiforme (Plate XXI. *p*) is conical with an obtuse apex, having **on its** base the articular surface for the os cuneiforme, and with the rest of its exterior surface more or less irregular, for implantation of a tendon and ligaments.

Trapezoides.—The homologue of the trapezium being connate with the scaphoid, and noticed in the description of that compound bone, the trapezoides (ib. *z*) is the first independent carpal bone of the distal row. It is the least of the carpal series, and is a relatively smaller and flatter bone than in the Mylodon: the proximal or scaphoidal **surface is** convex transversely, concave from behind forward, and plays in a corresponding concavo-convex surface in the scaphoid. The distal surface is almost wholly convex: **both** surfaces are joined by a small articular facet **on the** radial side of the bone, which is adapted to a corresponding facet in the small metacarpal bone of the thumb; and by **a more extended articular surface** on the ulnar side of the bone for junction with the os magnum.

Magnum.—This bone, arbitrarily so termed, comes next after the trapezoides and pisiforme, in the order of size, being much inferior to the other carpals. It is almost wholly covered by smooth articular surfaces. The small non-articular rough surface (ib. *g*) exposed upon the back of the wrist is of a transversely extended hexagonal figure, with the **outer and inner sides the shortest.** The surface for the lunare is concave anteriorly, but very convex in the greater part of its extent. It is continuous at its radial border, **with the two** surfaces, **one concave, the other flat,** for the scaphoides; and at its ulnar border **with the flat** surface for the unciforme. The concave scaphoidal surface is continuous with **the surface for the** trapezoides, which is much narrower than is the same **surface** in the Mylodon, the scaphoidal surface being broader than in the Mylodon. The largest articular surface is that for the base of the great middle metacarpal, which is slightly convex except at its fore-part, which is produced into a cone.

Unciforme.—Of all the carpals the os unciforme (ib. *u*) differs most in form from that of the Mylodon; it is a thick transversely extended bone, the free tuberous surface on the back of the wrist being hexagonal, with the two inner and the two outer sides very **short: one** of the outer sides is formed by a protuberance separating the articular sur**faces for the** os cuneiforme and fifth metacarpal, which meet at an acute angle in the Mylodon.

The proximal oblong concavo-convex surface for the cuneiforme is continuous, at the radial side of the bone, with the surfaces for the lunare and os magnum; and the latter with the broad surface along the distal part of the base which is obscurely divided into three facets, the middle and largest for the base of the fourth metacarpal, the next in size for **the** outer facet or the base **of** the third metacarpal, and the outermost and smallest facet for the small part of the base of the fifth metacarpal which articulates with the carpus. The inner or palmar rough surface of the unciforme has an oblong tuberosity, which is narrower than that upon the dorsal surface.

Metacarpus.—The innermost or first metacarpal (Plate XXI., 1, *m*), answering to that of the pollex or thumb, resolves, by its rough obtuse distal termination, as well as by its diminutive size, one of the points considered doubtful by CUVIER, in the structure of the fore-foot of the Megatherium, by proving that the pollex was absent, or represented solely by the rudimental metacarpal, which must have been concealed beneath the integument. The bone is of an irregular subcubical shape, broader than it is long. At its base are two separate subcircular flat surfaces articulating with corresponding facets on the trapezial part of the scapho-trapezium: on the side next the second metacarpal is a convex articular surface, having the lower part obscurely defined for articulation with the trapezoides, and the rest lodged in the concavity, partly articular, partly rough, upon the outside of the base of the second metacarpal. The outer side of the rudimental first metacarpal is obliquely flattened, the surface here being apparently for the insertion of a strong tendon.

The metacarpal of the second digit (ib. *m* II*) has a very irregular exterior; its comparatively small base is excavated obliquely to receive the fore-part of the trapezoides; on the outer or radial side it shows a triangular excavation, the lower half of which has a smooth articular surface for the rudimental metacarpal, *m* I. On the ulnar side of the base there is a convex articular surface divided into a proximal narrow tract for the os magnum, and a distal broader tract for the contiguous side of the base of the middle metacarpal. Above or beyond this the middle third of the bone supports a rugged protuberance, which has been attached by ligament to a similar rough surface on the middle metacarpal; so that little more than one-third of the bone projects freely from the metacarpus. This free part is subcompressed and expands in the direction from the back to the palm of the hand, so as to form the surface for a trochlear articulation of great extent in that sense. A ridge from the back surface of the metacarpal expands into a tuberosity near the dorsal end of that articulation; and a similar but smaller tuberosity projects from the palmar end of the same articulation, along each side of which there runs a thickish edge. The distal surface itself presents a median vertical or longitudinal prominence, beyond the lower half of which the articular surface is produced laterally so as to make the surface concave transversely on each side the median ridge; whilst at the upper half the whole surface is convex transversely, as it is, in a minor degree, longitudinally.

The middle metacarpal (ib. *m* III) is a little longer than the second, but is twice as thick, with a quadrate transverse section, the four sides of the shaft being flat and sharply defined. The base of the bone is produced 'ulnad' and 'proximad,' so as to be wedged between the fourth metacarpal, the magnum and the unciforme. The articular surface on the ulnar side of this production for the fourth metacarpal is extensive, and for the

---

* Referred to the 'annulaire' or fourth digit in the 'Ossemens Fossiles,' *ed. cit.* pl. 216, fig. 8.

most part slightly convex: the surface on what may be termed the truncate end of the production, for articulation with the unciforme, is slightly convex on the dorsal half, slightly concave on the palmar half. The surface for the magnum on the proper base of the metacarpal sinks into a conical cavity near its dorsal end; the rest being nearly flat. On the radial side of the base is an oblong articular tract for the second metacarpal, and beyond this the extensive rough surface for the ligamentous connexion with the same bone. The radial surface is grooved above the rough facet, obliquely by a wide and moderately deep canal; apparently for the passage of a tendon to the digit. The dorsal surface shows an oblique broad tuberosity, extending from above the tendinal groove to the upper or dorsal end of the distal articulation. The smooth and almost flat dorsal surface gradually deepens into a broad and shallow oblique channel on the ulnar side of the oblique tuberosity. The ulnar side of the bone, beyond the articular surface for the fourth metacarpal, is occupied by a rugged flat tract for ligamentous connexion with the same bone. The palmar surface is pretty smooth, flat transversely, slightly concave lengthwise; produced into a tubercle below the middle prominence of the distal joint. The articular surface of this joint does not cover the whole distal end of the bone; it is long and rather narrow, extending obliquely from the palmar forward to the dorsal surface; the dorsal side of the bone being longer than the palmar one. It is traversed lengthwise by a median prominence, convex transversely, almost straight lengthwise; and the surface is continued upon a flat tract on each side of the prominence, that on the radial side of the prominence being the broadest.

The general form of this bone is more like that of a brick or of an ashlar stone for a strong wall, than like that of the usual support of a flexible digit of a fore-paw or hand.

The chief difference between the middle metacarpals of the Megatherium and Mylodon is in the form of the distal articulation. This surface, in the smaller Megatherioid, is convex from above downward, whilst in the Megatherium it is straight, or rather concave, and it joins the dorsal surface at an acute angle. The lateral depressions of the pulley are narrower in the Megatherium, and the vertical inflexions of the phalanx must have been more limited than in the Mylodon.

The fourth metacarpal (ib. m IV), as compared with the third, is longer and more slender in the Megatherium than in the Mylodon; but its articulation by an obliquely extended base with the third and fifth metacarpals and the unciform bone, closely corresponds with that in the Mylodon*.

The two oblique metacarpal surfaces are nearly parallel, the radial one is concave, the ulnar one slightly convex; both are separated by a sharp angle from the intermediate or carpal surface, which is nearly square and is slightly concave. The proximal half of the bone is bounded by four flat equal sides, with intervening angles, sharp on the radial side. The upper and under sides are nearly smooth, with a broad low tuberosity near the proximal end; the outer and inner sides are rugged for close syndesmosis with the adjoining metacarpals. Only the distal half of the bone stands freely out; it expands

* Description of the *Mylodon robustus*, 4to, p. 92. pl. xv. *n* 4.

in vertical breadth to the articular surface for the fourth finger. The angle between the upper and radial sides is rounded off; that between the upper and ulnar sides is, in one specimen, developed into a sharp ridge.

The distal articular surface resembles that of the second metacarpal, but **occupies a** smaller relative proportion of **the whole distal surface**; the vertical prominence, convex transversely, is broader, more concave vertically, and extends obliquely from the upper **and** radial angle to the lower and ulnar one: the flat lateral extension of the articular surface is confined to **the** radial side of the prominence. A single flat surface for a sesamoid bone is situated below, but distinct from, this part of the articulation. A large and strong **tuberosity projects below the** prominent part of the joint; an oblique channel **divides this tuberosity** from a longer one on the ulnar side of the distal end of the metacarpal.

The distal **articular surface of the** fourth metacarpal **in the** Mylodon is reduced to a small, vertical, oblong, nearly flat surface; this difference relating to **the stunted development** and limited function of the digit it has to support. In the Megatherium such simplification of the distal joint of the metacarpal is limited to the fifth of that series of bones (ib. *m* v): this metacarpal *, of the same length as the fourth, is more slender; its proximal end is wedge-shaped, the radial articular surface and the flattened outer facet converging to the narrow rough tract which is joined **by ligament** to the carpus. The articular **surface on** the radial side has a **small terminal part obscurely marked off for a facet on the unciforme; the rest receives the convexity on the ulnar** side of the fourth metacarpal: beyond this is a rough surface for the syndesmotic union of the contiguous bones. Rather more than half the fifth metacarpal stands freely out; it is traversed above by a longitudinal ridge expanding into a broad tuberosity at the distal end; the **bone is smooth** and rounded on the palmar side. The small terminal articular surface **is oval,** slightly concave vertically, concave transversely, upon the radial half of the distal expansion. **There is no** articular surface for a sesamoid.

The metacarpals are so united and wedged together and with the carpus, as to transmit from the oblique carpal surface which sustains the radius the weight of the fore part of the Megatherium to the fifth digit, the stunted extremity of which was imbedded in the marginal hoof-like callosity on which the ponderous quadruped trod, with the claw-bearing toes bent inwards.

From the scapho-trapezium and lunare the weight was transmitted to the second row of carpal bones; and by **the** oblique production of the base of the second metacarpal, and especially of the third metacarpal, it was concentrated through the medium of the fourth metacarpal upon the fifth. The lateral pressure thus occasioned explains the extent of syndesmotic and sutural union along the basal part of the metacarpals, and also the squared angular shape of the constituents of this masonry of the huge fore-paw.

On comparing the structure of the carpus in the Megatherium with that in existing Mammals, it is found to be repeated in the Unau or Two-toed Sloth (*Bradypus* (*Chola-*

* Referred to the index or second digit in the 'Ossemens Fossiles,' *ed. cit.* pl. 16. fig. 11.

pus) didactylus), and in that species or subgenus exclusively; the carpal bones being seven, and the reduction to that number resulting, also, from the connation of the **scaphoid and trapezium**. A scapho-trapezial bone exists in the Aï or Three-toed Sloth (*Bradypus* (*Acheus*) *tridactylus*), but the carpus is further reduced in this species or subgenus to **six bones** by the confluence of the trapezoides with **the os** magnum. The trapezius is a distinct ossicle in the Chlamyphorus, as in **most other** Armadillos; in the *Dasypus sex-cinctus* it coalesces with the trapezoides. In the Pangolins (*Manis*) the scaphoid coalesces with the lunare, not with the trapezium. In the true Anteaters (*Myrmecophaga*), and in Orycteropus, the ordinary eight carpal bones retain **their individuality**.

Digital Phalanges.—The stunted metacarpal of the pollex, in the Megatherium, bore no rudiment of a digit. They were powerfully developed and **unguiculate in the three following digits.**

The index digit (Plate XXI., ɪɪ) has three phalanges: the proximal one (1) is almost twice as broad, and more than twice as deep, as it is long; the metacarpal surface presents a deep and wide, vertically elongated, subangular concavity, fitting the vertical prominence and lateral facets of the distal joint of the metacarpal; the distal surface of the proximal phalanx presents a vertical angular fissure dividing two oblong convexities. The mid-phalanx (*ib.* ɪɪ, 2) has a proximal trochlea playing on the preceding, the median vertical ridge being overhung by the produced upper rough surface; the distal articulation repeats that of the proximal phalanx, but **the median fissure is less deep, and the** lateral convexities are more regularly rounded and prominent. The **ungual phalanx** (*ib.* ɪɪ, 3) exceeds the length of both preceding phalanges; the upper or dorsal **side of its base is produced backwards into an obtuse point; the rough sheath of the claw extends along three-fourths of the phalanx;** it is convex radiad, vertical and flat ulnad; the core presents **a rough edge radiad and** a vertical rough surface ulnad, and is smooth and convex transversely **at its** base, both above and below.

The proximal and **middle** phalanges of the digitus medius (*ib.* ɪɪɪ, 1..2) are confluent, the line of anchylosis being indicated by a vertical ridge along both the inner and the outer sides, and by a curved ridge convex backwards on the upper or dorsal side. The proximal articular surface **presents a** deep vertical channel, with a narrow elongated subconcave surface **continued from its** radial side, and a still narrower flat surface from the opposite side. The distal articular surface of the composite bone has a deep median vertical groove dividing two convexities. The compound phalanx presents two rough tuberosities below, one terminating each **bank of** the vertical articular proximal channel; there is a smooth depression above, and another below, close to the distal trochlea. The enormous distal phalanx (ib. ɪɪɪ, 3), of twice the vertical breadth of the adjoining claw phalanges, is very little longer than that of the second digit; it is flattest on the ulnar side, to which the upper convexity slightly inclines. The base or palmar part of the phalanx is the broadest, is convex both lengthwise and transversely, and is perforated by **two** canals, for the large vessels and **nerves** supplying the claw-core. The surface, in which was implanted the flexor tendon, shows many linear impressions radiating from the median tuberosity, forward and laterally. The articular surface excavates obliquely the base of this phalanx which overhangs the joint; the concave border of the median angular prominence describes a semicircle from above forward and downward; the ulnar excavation is longer in that direction than the radial one. The well-marked trochlear joint restricts the movements of the great claw to flexion and extension on a vertical plane, whilst its position effectually prevents retraction of the claw, the preservation of the effective condition of which is due to an opposite bend to that in the Cat-tribe, as was well explained by CUVIER in his description of the Megalonyx*. The major part of this enormous phalanx goes to form the sheath of bone protecting the base of the claw. The bony core or peg on which the claw was fixed and moulded is compressed, with a sharp edge above and flat below, where it projects beyond the sheath.

The finger of the fourth digit has three phalanges and is unguiculate, with a claw resembling in shape and size that of the second digit. The proximal phalanx (ib. IV, 1) is so compressed in the direction of its axis, that the proximal and distal articular surfaces almost meet above, only about $1\frac{1}{2}$ line's breadth of rough surface there intervening: the proximal articular surface is a wide and moderately deep vertically-elongated channel, with a semielliptic flat surface continued from the lower half of the radial border; the outer and inner sides of the phalanx are narrow, elliptic rough convexities; the distal articular surface is gently convex vertically, sinuous laterally. The second phalanx (ib. IV, 2) is deeper than long; its upper surface is produced backward over the proximal phalanx into a rough obtuse process; the under surface is much shorter, but is broader than the upper surface; the convexity of the distal articular surface describes a semicircle from above downward, and is divided by a vertical trochlear groove into two parts, of which the ulnar one is the larger. The ungual phalanx (ib. IV, 3) is long, triedral, with the radial side of the sheath vertical, the under surface rough and flattened, and the ulnar surface sloping from above downward and outward to join the under surface. The upper posterior tuberosity, overhanging the joint, inclines radiad, and is somewhat flattened; the proximal surface presents a median vertical rising for the channel in the preceding joint.

The small elliptical surface on the radial half of the distal end of the fifth metacarpal supports a very short lamelliform phalanx, produced outward some way beyond the joint; and to this is articulated a stunted second phalanx with a rough, obtuse, non-articular end. In some instances the above two phalanges are blended together.

Amongst existing Mammals the Sloths alone present the connation of the scaphoid with the trapezium, the Two-toed Anteater and the Armadillo that of the first with the second phalanx. This latter character, peculiar to the middle digit in the Megatherium, is limited also to the same digit in the *Myrmecophaga didactyla*; in *Dasypus* it affects the third, fourth and fifth digits.

* Ossemens Fossiles, ed. cit. t. viii. p. 510.

The *Bradypus tridactylus* has but two flexible phalanges in each of the three unguiculate toes, but this reduction is due to the early anchylosis of the proximal phalanx with the metacarpal, not, as in Megatherium, with the second phalanx. In the *Bradypus didactylus* the unguiculate digits preserve the normal number of free phalanges.

The pollex is atrophied in the Megatherium, as it is in both existing species of Sloth; and, as in the *Bradypus tridactylus*, only the second, third and fourth digits support claws; but the fifth digit, instead of being wanting, as in the Ai, is developed, so far as was needed for the purposes of terrestrial progression, in the Megatherium. The small existing arboreal Sloths are seldom obliged to walk on the ground, and there can only crawl along with difficulty. In the Mylodon the pollex was developed and unguiculate, but both the fourth and fifth digits were terminated by a stunted second phalanx. In the Unau, not only the fourth and fifth digits, but also the first are suppressed in the forefoot. Yet this is the Sloth, as already remarked, which so peculiarly illustrates the bradypodal affinities of the Megatherium in the structure of the carpus, notwithstanding the degree to which the adaptive modifications of the Megatherioid type of fore-foot are carried in relation to the exclusively arboreal life of this small existing tardigrade. The coalescence of the scaphoid and trapezium, which Cuvier was the first to recognize in the existing Sloths, he continued to affirm in the latest edition of the 'Ossemens Fossiles' to be peculiar to them. The bony structure of the fore-foot of the Megatherium he regarded as most resembling that of the *Dasypus gigas*. M. Laurillard, after the subsequent reception of casts of the carpal bones of the Megatherium, which had been transmitted to England, with other bones of the Megatherium, by Sir Woodbine Parish, K.H., inferred that the fore-foot of the Megatherium had a greater analogy with that of the *Myrmecophaga jubata*, but he did not detect the connation of the scaphoid with the trapezium. The form of the scapho-trapezial bone in both existing Sloths bears an unmistakeable resemblance to that in the Megatherium, but in the Unau it describes a deeper curve towards the palmar aspect, and the trapezial portion (described by De Blainville as the sesamoid of the pollex[*]) is relatively longer than in the Megatherium. The base of the stunted metacarpal of the pollex is expanded, and abuts by one part against the trapezium and by another against the base of the second metacarpal. The trapezoides is a small bone articulated, as in the Megatherium, with the scapho-trapezial, the os magnum, and the second metacarpal; the os magnum presents almost the same pentagonal contour, dorsally, as in the Megatherium, the anterior facet being also partly convex for adaptation to a concavity in the base of the middle metacarpal, which likewise is so extended as to interpose itself between the fourth metacarpal and the carpus.

The atrophy of the fifth finger, which has proceeded in the Megatherium to cause the absence of the ungual phalanx, and which atrophy similarly affects both the fifth and fourth digits in the Mylodon and Scelidotherium, has proceeded in the

[*] Ostéographie de Paresseux, 4to, p. 22.

Unau to the removal of all the bones of those digits, save the metacarpal of the fourth, which is reduced to a rudiment of even smaller size than that which forms the vestige of the thumb on the radial side of the hand: it rests, as a great part of the fourth metacarpal in the Megatherium does, upon the expanded base of the third metacarpal.

In the Ai (*Bradypus tridactylus*) the metacarpal rudiment of the pollex is anchylosed at its lateral joint to the base of the metacarpal of the index, but it retains its free articulation with the scapho-trapezium. The chief modifications of both hand and foot in the Three-toed Sloth are the extensive anchyloses of different bones: this character is shown by the coalescence of the trapezoides with the os magnum, such compound bone supporting the base of the second metacarpal and a great part of that of the middle metacarpal; thus fulfilling the same relations to the metacarpus as do the separated bones in the Unau and Megatherium.

The great extent to which the metacarpals are suturally united to each other in the Megatherium, is a character repeated in those of the Ai, but the suture is speedily, in the living Sloth, converted into bony union, and the three metacarpals, like the three metatarsals, thus form one compound bone, as in Birds. The unguiculate digits which this bone supports in the fore-paw, are the homologues of the three claw-bearing toes in the Megatherium. The rudiment of the fifth finger appears as a mere process from the outside of the base of the metacarpal of the fourth: the huge terrestrial predecessor of the small leaf-eating and tree-dwelling quadrupeds retained the fifth toe, minus its terminal phalanx, yet of great size and strength, and modified expressly for the purpose of supporting the ponderous body in terrestrial progression.

The fore-foot of the Mylodon more closely conforms, in its essentials, to the type of that of the Unau, inasmuch as the two outer digits (fourth and fifth) were mutilated and clawless; they were, however, developed to the same degree as the fifth digit is in the Megatherium, and for the same end, but probably made little show, externally, in the entire foot. The pollex, however, instead of being rudimental, was fully developed, though small, in the Mylodon. In the Megatherium this digit is rudimental, as in both forms of existing Sloth; but the bones of the fore-foot correspond more closely with the type of the manus in the Ai; there being, indeed, as regards the digits, only this essential difference, that the fifth, instead of being, like the first, a mere rudiment, was developed to be adapted to progression on the ground. It is most interesting, however, to trace the interchangeable relations between the two above-cited great extinct Megatherioids and the two existing forms of Sloth, respectively.

In regard to other existing Edentata, the *Myrmecophaga jubata*, by reason of the clawless condition of its fifth digit, and the *Myrmecophaga didactyla*, by that of the rudimental pollex as well as fourth and fifth digits, ought to succeed the Sloths as next of kin to the Megatherioid quadrupeds, the interval being due to the difference of carpal structure.

Cuvier has observed that the fore-foot of the *Dasypus gigas* is one of the most extraordinary among quadrupeds; and, he adds, that it alone would give the key to all the anomalies in that of the Megatherium*. But this could only have been affirmed under a misconception of the real nature of those anomalies. In the *Dasypus gigas* the forefoot is pentadactyle; all the digits are unguiculate, and, in three of them, the **claw-phalanges** furnish a bony sheath as well as core to the claws; but these belong to the third, fourth and fifth toes, not, as in the Mylodon, to the first, second and third, or, as in the Megatherium, to the second, third and fourth toes; they moreover successively decrease in size from the radial to the ulnar aspect, instead of the reverse proportions which they present in the Mylodon. No doubt the claw on the middle digit is the most developed, as in the Megatherium, and the first and second phalanges of **this digit** have coalesced; but here ends the particular resemblance between the Megatherium and the great Armadillo, in regard to the bony structure of the fore-foot.

We can state with confidence, what M. LAURILLARD suggests†, viz. that the fore-feet of the Megatherium, as represented by the skeleton in the Madrid Museum, are not transposed, the right being on the left and the left on the right side, as CUVIER was led to suspect; but that the articulation of those complex parts by the laborious Prosector and Curator BRU, was in the main correct.

The bony structure of the fore-limb of the Megatherium is now, indeed, as completely understood, and the homologies of every constituent bone can be as exactly defined, as in any existing species of quadruped. And, to the degree in which so important a part of the frame throws light on the whole, the Naturalist may **thereby trace the affinities**, and the Physiologist infer the **habits, of the great extinct beast.**

§ 8. *Of the Bones of the Posterior Extremity.*

In the description of the bones of the hind-limbs I commence with the ilium, as being the homotype or correlative of the scapula in the fore-limb. The ischium, which is the homotype of the coracoid, is confluent with the ilium, as the coracoid is with the scapula; the pubis, which is the homotype of the clavicle, is confluent with both the ilium and ischium. All the three bones on both sides become confluent with the sacrum, and form therewith, in the full-grown Megatherium, a single bone—the pelvis—which is the largest known among terrestrial Mammals, and forms the most striking feature in the skeleton of the present gigantic extinct Sloth.

As a like progressive ossification brings to pass a similar state of the pelvis in the small existing Sloths, the limits of the primitive bony constituents of that of the Megatherium can only be determined, **at present, by** analogy with the pelvis of its modern congeners, studied at a period of immaturity. **This I** have had the opportunity of doing in the

---

* " La main du *tatou géant* est une des plus extraordinaires qu'il y ait parmi les quadrupèdes, et à elle seule elle expliquerait toutes les anomalies que nous verrons dans celle du Mégathérium."—Ossemens Fossiles, *ed. cit.* tom. viii. p. 242. No qualifying note is appended by the editors to this statement.

† See the note (1) appended by that able anatomist to the chapter on the Megatherium, in the posthumous edition of the 'Ossemens Fossiles,' 8vo, t. viii. p. 355.

young of the *Bradypus tridactylus*, prior to the completion of the coalescence of the several bones.

The five vertebræ composing the sacral part of the pelvis of the Megatherium have been described in § 2, p. 22, treating of the vertebral column to which they belong.

The iliac bones (Plates I. and XXII. *a*), as they extend from their place of anchylosis with the sacrum, expand in depth and breadth; their anterior plane is directed forward, being almost vertical and at right angles with the axis of the spine. Each ilium, after contributing its share to the acetabulum (Plate XXII. fig. 1, *a*), rapidly contracts to an obtuse point bent downward and outward. The two bones, in a front view, resemble a pair of broad outstretched wings at the sides of the fore part of the sacrum. The anterior surface is slightly concave, but is undulated, with many sharp ridges that have penetrated between the fasciculi of the muscles thereto attached. The 'labrum,' or upper and outer convex border of the ilium, is unusually thick and rugged; the under concave border is also rugged, but is thin, and in some parts sharp. On the inner side of the acetabulum there is a well-defined, raised and very rough, oblong surface (*p*) for the insertion of the tendon of a powerful 'psoas' muscle.

The outer surface of the ilium is slightly concave near the sacrum, and is then convex in the direction of its longest diameter, which is from within outwards; in other directions it is nearly flat: its surface is much broken by numerous intermuscular ridges.

The pubis (Plates I., XXII. & XXVII. *a*) is very slender where it forms the anterior border of the 'foramen ovale' (Plate XXII. fig. 1, *o*), but expands at its extremities, and especially where it coalesces with the ischium to form the produced and pointed 'symphysis.' The extent of this symphysis is 10 inches in a straight line.

The pelvis, as in the Sloths and other members of the Order BRUTA, shows the conversion of the ischiadic notches (ib. *i*) into foramina by the anchylosis of the ischia with the posterior sacral vertebræ (ib. *s*). Each ischium (Plates I. & XXII. *a*), as it extends from its confluence with the sacrum, expands into a broad smooth plate of bone, bent outward and forward, then contracting as it converges inward towards its fellow, to combine with the pubic bones at the symphysis. The hinder border, forming the tuberosity, *a*, is thick and rugged; and two or three perforations here indicate the original line of its separation from the sacrum. The part of the ischium which joins the pubis on the sacral side of the foramen ovale presents on its inner surface the usual oblique channel leading to that foramen.

Among the existing species of the Order BRUTA the Sloths alone resemble the Megatherium in the expansion of the iliac bones, but this is much less in comparison with the length of the trunk; the iliac expansion is relatively greater in the Megatherium than in the Elephant, and is associated with a much greater proportionate size of the whole pelvis and of its cavity or channel. The extreme breadth of the pelvis of a large Asiatic Elephant is 3 feet 8 inches, whilst in the Megatherium it is upwards of 5 feet.

The pelvis which CUVIER was led to suspect, from the defective condition of its fore part in the Madrid skeleton, to be naturally open anteriorly, as in the *Myrmeco-*

phaga didactyla, is closed anteriorly, as in the Sloths, by a 'symphysis pubis' of short extent.

The acetabulum (Plate XXII. fig. 2) presents a full oval shape, with the lower margin bisected by a narrow and deep 'Haversian' groove, which extends, slightly expanding and becoming more shallow, to near the bottom of the cavity; the outer division of the lower border of the groove is most produced. The large and deep acetabula look downward and a little outward. One diameter of the hemispheroid cavity is 8 inches, the other diameter is 7 inches; it therefore presents a plane surface of 43·9824 square inches, which, multiplied by 15, with the barometer at 30 inches, gives about 660 pounds atmospheric pressure upon the hip-joint of the Megatherium.

The size and strength of the ordinary processes of the pelvis, the breadth of the rough labrum of the iliac bones, and the numerous and well-defined intermuscular crests, indicate the unusual size and vigour of the muscular masses which proceeded from the pelvis in different directions to act upon the trunk and fore limbs and upon the hind limbs and tail. They lead to the conviction that the resistance which demanded such forces for its overcoming must have been of a very different nature and degree from any that now opposes itself to the labours of the existing vegetable feeders when engaged in supplying their daily wants; whence it may be inferred that the exertion of such forces was associated with equally peculiar habits in the megatherioid animals.

The femur of the Megatherium (Plates I. & XXVII. 6s, and Plate XXIII. fig. 1) is one **of the most** massive limb-bones in the Mammalian class; from its proportions it might rank with the 'flat' instead of the 'long' bones, but that its thickness **would rather** bring it into the category in which the carpal or tarsal bones are placed according to the old anatomical character derived from shape.

The head (Plate XXIV. fig. 1) would be a smooth hemisphere, but that the anteroposterior diameter somewhat exceeds the transverse one. Its surface is unimpressed, and its periphery uninterrupted, save by a small entering notch at the middle of its back part (Plate XXIII. fig. 1), into which possibly some thickened band of the capsule, like a '*ligamentum teres*,' may have been implanted. The neck of the femur is short and ill-defined; the upper contour passes from the head to the summit of the great trochanter (**ib.** *t*), which is on the same level, in a slightly concave line; the inner contour of the bone descends with a somewhat deeper concavity from the lower periphery of the head to the shaft. This is flattened from before backward, and presents a slightly oblique twist, the head and the outer condyle being on a plane anterior to the trochanter and inner condyle. The great trochanter presents a broad rugged surface, flattened obliquely from above downward and inward, divided into two somewhat flattened facets by a transverse ridge arching upward. The lower facet contracts as it descends, and is continued into the strong outer ridge which descends to the external condyle. The thick rough border of the rest of the trochanter stands out beyond the contiguous parts of the femur; least so at the upper, and most at the back part of the process, where the border defines outwardly a small but deep trochanterian fossa (ib. f). This fossa is bounded below by a small tuberosity. The small trochanter is represented by a rough ridge 6 inches long, 2 inches broad, occupying the middle third of the shaft a little anterior to its inner border. A few shorter longitudinal ridges occur on the fore part of the shaft between the upper end of the small, and the upper and fore part of the great trochanter. The rest of the anterior surface of the femur is smooth, concave lengthwise, slightly convex transversely.

The back part of the femur presents a small low tuberosity below its middle part, near the inner border; the triangular surface between this tuberosity, the head, and the great trochanter, is smooth and flat. The contour of the back part of the femur, from the head to the outer condyle, is convex; that from the trochanter to the inner condyle is concave; the lower half of the back part of the shaft is convex transversely. The lower end of the femur (Plate XXIV. fig. 2) presents two articular surfaces, the inner one (ib. i) being that of the internal condyle, the outer one being the combined ectocondyloid (e) and rotular (r) surfaces. The latter is extensive, and describes a semicircle from before backward, but is narrow from side to side: in this direction the rotular portion is slightly concave; its limits are indicated by a notch on the inner side: the condyloid portion is slightly convex transversely, in which direction the extent is scarcely $2\frac{1}{2}$ inches. The entocondyloid surface (i) is of nearly twice that breadth; is of a reniform shape, and is convex in every direction. The intercondyloid channel is roughened by decussating ligamentous impressions and ridges, with some vascular pits. Its narrowest part is anterior, and measures an inch and a half across; posteriorly it is 3 inches across. The intercondyloid border of the inner condyle is sharply defined and projects below the level of the canal, that of the outer condyle is mostly on a level with the canal. The surface of the canal meets the hind surface of the femur at almost a right angle, the intervening ridge being rounded off. The fore part of the inner condyle stops short of the fore part of the shaft; the back part of the condyle projects 1 inch beyond the back part of the shaft. The process or prominence above the inner condyle (*eminentia entocondyloidea*, Plate XXIII. fig. 1, *ie*) forms an obtuse angle, the lower side of which is rough, rather flattened, and expands as it descends towards the articular part of the condyle. The outer condyle does not project so far back as the inner one: its supracondyloid prominence (*ec*) is larger, of a similar angular form, but is bent forward as well as outward.

In the Elephant, Mastodon, and Diprotodon, the shaft of the femur is flattened from before backward; but the length of the femur so far exceeds its breadth, that, strong as the thigh-bones of these quadrupeds are and well-proportioned to the weight they had to sustain, they appear weak and even slender when placed by the side of the femur of the Megatherium. The Rhinoceros, which has the thigh-bone relatively broader and flatter than in the proboscidian pachyderms, differs more markedly from the Megatherium in the presence of the third trochanter.

The shaft of the femur is more or less flattened in all the species of *Bruta*; but the Orycteropus and Armadillos, in which this character is conspicuous, differ, like the Rhinoceros, from the Megatherium in having the third trochanter. This process is not present in the Pangolins (*Manis*), Anteaters (*Myrmecophaga*) or Sloths (*Bradypus*); but in all these genera the femur is relatively longer and more slender than in the Megatherium, and only the Sloths amongst existing Mammals, not marine, repeat the remarkable megatherioid character of the absence of a medullary cavity in the shaft of the femur.

The general characters of the femur of the Megatherium are most closely repeated in that of the Mylodon* and Scelidotherium†. In these genera, however, instead of the shallow notch, there is a deep and prolonged fossa for the ligamentum teres on the middle of the hind border of the head of the bone. The post-trochanterian depression is relatively larger in the Mylodon, the small trochanter is relatively less and higher placed; the whole femur is longer in proportion to its breadth, and the distal expansion is relatively less. Here, also, the articular surfaces offer a well-marked character of distinction; the rotular articular surface is continuous with that of both condyles, and unites them anteriorly.

From the cast of a distal epiphysis, transmitted by Dr. HARLAN to the Museum of the Royal College of Surgeons, London, as of the Megalonyx, it would appear that this extinct megatherioid offered a third modification of the knee-joint, the rotular surface being distinct from those of both condyles. Thus the knee-joint of the Mylodon must have had one large synovial capsule, that of the Megatherium two, and that of the Megalonyx three such sacs.

The patella (Plate I. *o*, and Plate XXIII. fig. 2) is a strong, thick, subtrihedral conical bone, with the base rounded, and also the angle between the two rough outer sides. In the natural position of the bone the base is uppermost, and chiefly composed of a strong tuberosity coarsely and irregularly striated: at the lower half of the outer surface the striæ have a longitudinal and subparallel direction, giving that part of the bone the appearance of an ossified fibrous ligament. On the inner and broadest side the articular surface occupies the upper two-thirds: it is less distinctly divided by a median longitudinal rising into two channels than in the Mylodon, being more nearly level and uniform. The non-articular surface below the joint is irregularly grooved and perforated by vascular canals.

The fabella‡, or post-tibial sesamoid bone (Plate I. *v*), is a smaller subhemispheric bone, with a circular, slightly concave articular surface, which was applied to part of the outer condyle of the femur: the rest of the surface is rough and fibrous, indicative of the imbedding of the bone in a flexor tendon of the leg.

The tibia and fibula (Plates I. and XXVII. *o, v*) become anchylosed together at both extremities in the Megatherium: they are both short, and the tibia presents massive

* OWEN 'On the *Mylodon robustus*,' 4to. p. 111. pl. 17.
† Ib. 'Fossil Mammalia' of the 'Voyage of the Beagle,' 4to. pl. 25. fig. 5.
‡ Ib. 'Archetype of the Vertebrate Skeleton,' 8vo. p. 190.

proportions corresponding with those of the femur. The proximal end of the tibia (Plate XXIV. fig. 3) presents two distinct and well-marked articular surfaces; the inner one (*i*) is concave, the outer one (*e*) is convex: the extent of these surfaces corresponds with the breadth of the articular part of the outer and inner femoral condyles respectively. The back part of the outer facet which bends downward affords an articulation (*f*) to the fabella. The rough interspace between the articular surfaces is a little concave transversely, and convex from before backward; its breadth equals that of the outer surface: it developes no intercondyloid process for crucial ligaments.

The fore part of the proximal end of the tibia presents a large irregularly triangular rugged protuberance for the ligamentum patellæ; the back part, below the outer condyle, developes a smaller but more prominent rugged process, the apex of which overhangs the upper part of the interosseous space: between this 'post-tibial' process and the rough inner border of the bone there is a deep and wide longitudinal channel inclining a little obliquely to the interosseous space. The shaft of the tibia gradually contracts to its middle, and as gradually expands to its distal end. It is subcompressed from before backward; is smoother behind than in front: there is a longitudinal channel on each side the back part of the lower end of the tibia, which forms a convexity between them.

The anterior surface is divided by a ridge extending obliquely from the rotular protuberance to the inner malleolus: the surface on the fibular side of this ridge is smoother than the other, which seems to have been wholly given up to musculo-tendinous attachments. The inner malleolus (Plate XXIV. fig. 4, *m*) is a slight expansion below the confluence of the inner and oblique anterior ridges: it does not project below the level of the distal articular surface. This is deeply concave, and is divided into two facets by the deeper hemispherical excavation near its inner side for the reception of the inner protuberance of the astragalus: this excavation (*e*) gives to the larger and shallower facet (*i*) a full crescentic figure. The smooth surface is sometimes continued upon, sometimes interrupted by, a narrow tract from the vertical surface upon the malleolar end of the fibula, which surface is applied to the outer facet of the astragalus. There is a small orifice for a medullary artery at the middle of the back part of the tibia, but it does not open into any medullary cavity: the bone is cancellous throughout.

The fibula is thickest at its upper end, where it has a trihedral form; the outer surface is convex and rough, the inner and hinder surfaces are concave and smooth, meeting at a sharp interosseous border directed obliquely backward: this border is slightly thickened and produced at its middle part, where the shaft of the fibula is compressed: it augments in thickness and resumes its trihedral shape as it descends, and terminates in a moderately produced outer malleolus (*f*) with a very rugged surface, except where it articulates with the astragalus.

In the Mylodon, as well as in the Megalonyx and Scelidotherium, the tibia and fibula continue separate,—a fact affecting the value of the evidence which CUVIER deduced from their anchylosed condition in the Megatherium in favour of its affinities to the Armadillos, to which this structure is peculiar amongst existing Mammals. But, since it is known only in the Order *Bruta*, it forms an interesting additional proof of the essential relations of the huge extinct animal under description to that now anomalous group of Mammals. The tibia of the Mylodon is proportionally shorter and thicker than in the Megatherium: the outer articular surface at the upper end is of a subcircular form and slightly concave: in the Megalonyx it is convex, but in a less degree than in the Megatherium. The lower articular surface in both Mylodon, Scelidotherium, and Megalonyx, presents the additional facet for the fibula. The hemispheric excavation on the inner side of the distal articulation is relatively larger in the Mylodon than in the Megatherium. This excavation, with the concomitant protuberance of the astragalus, is peculiar to the great extinct Sloth-like quadrupeds; in which so secure an interlocking of the foot with the leg bespeaks some habits peculiar to them, connected with the requirement of unusual resistance in the foot to the forces acting upon it from the leg and thigh.

In the series of existing animals Man presents the plantigrade foot in which the weight of the body presses most nearly upon the crown of the tarsal arch; but, in the Megatherium, owing to the length of the heel and the shortness of the toes, the leg transmits the superincumbent weight nearly upon the middle of the foot. So singularly shaped and adjusted, however, are the tarsal bones in the Megatherium, that the tibia articulates with the side instead of the summit of the tarsus, so that the whole foot is turned inward and rests upon its outer edge instead of its sole (Plate XXVI. fig. 1, and Plate I. *cl*, *a*).

The number of tarsal bones is reduced to six, through the absence of the entocuneiform*; but the astragalus, and especially the calcaneum, are developed to a great size.

The astragalus (Plates XXV. & XXVI. fig. 1, *a*) is of a peculiarly irregular form: if the foot be placed with the sole flat on the ground, as in Plate XXV., the chief articular surface (*a*) for the tibia looks inward, and the small fibular facet, at right angles therewith, is uppermost. The extensive surface by which it articulates with the bones of the leg is divided into three parts, the planes of which are at right angles to each other. The middle and largest division (*a*), answering to the outer ridge of the trochlear surface in the common form of astragalus, is here expanded into a broad reniform smooth tract, horizontal in the ordinary position of the Megatherium's foot (Plate XXVI. fig. 1, *a*), almost flat from before backward, convex from side to side. This surface is continued over the outer edge upon the outer side of the bone, in a triangular form, with the apex rounded off, to form the facet (*o*) for the fibula. The surface answering to the main part of the trochlea and its inner malleolar facet in ordinary astragali, is here reduced to a small triangular convexity (*ib.* & Plate XXV. fig. 1, *i*), forming the third **and** internal division of the surface, and supported on what appears to be an obtuse pyramidal process from the inner and lower part of the bone. This convexity is wedged

* 'Os cuneiforme internum' of SŒMMERRING.

into the deeper excavation on the inner part of the tibial articular surface, and forms a kind of pivot on which the foot worked.

The under or calcaneal side of the astragalus slopes from behind downward and forward, and is divided by an oblique groove, about an inch broad, into two facets; **the outer and posterior one is for the calcaneum exclusively; the inner and anterior one is for the calcaneum**, but is continuous with the surface for the naviculare. The outer calcaneal surface is ovate, concave lengthwise, convex across its posterior broader end, nearly flat at its anterior end. **The inner calcaneal** surface is of much less extent and **is nearly flat; its anterior** end suddenly bends forward and upward to be continued into **the outer convex part** of the navicular surface, which surface is divided into this convex portion and a contracted **subcircular concavity. The** lower part of the convex facet rests upon an **articular concave surface on the cuboides.**

Only a small proportion of the outer surface of the astragalus is non-articular, and this is chiefly on the outer, or, in the megatherian position of the foot, the upper surface (Plate XXVI. fig. 1). The groove dividing the calcaneal surfaces begins at their lower part and passes backward and inward, with slight terminal bends, which give it the form of the italic f. It is divided by a ridge from a wider channel, excavating the under part of the inner process; from which a third channel extends backward, dividing part of the inner calcaneal facet from the **naviculo-cuboid surface.** Both this channel and the calcaneal one are perforated by conspicuous vascular canals.

The calcaneum (Plates XXV. & XXVI. fig. 1, *b*) **is a long irregular pyramidal bone, with an obtuse** apex: the base is obliquely **truncated, forms the fore part of the bone, and** supports two articular surfaces (Plate XXVI. fig. 2): **the upper and outer surface (*as*) is ovate,** concave where **that** on the astragalus is convex, and *vice versâ*: **the oblique channel which divides this astragalar surface from the** astragalo-cuboidal surface (*as'*) is **as deep as** in the astragalus, and, like it, is perforated by vascular canals. The slightly **concave** triangular inner astragalar facet (*as'*) is continued below, at an obtuse angle, into the semielliptic slightly concave surface (*cb*) for the os cuboides. External to the astragalar **facet a strong** vertically extended tuberosity (Plate XXVI. fig. 1, *t*) forms the fore part of a wide and deep tendinal canal (*u*); a narrow **and feeble ridge bounds it** behind; a second and more shallow groove (*v*) succeeds; and, behind this, is a wide and deep vascular perforation (*w*). **The** under part of the calcaneum is very rugged and flat. The obtuse hinder extremity of **the** heel-bone shows by its sculpturing **and** the outstanding osseous spiculæ, the force with which the attached cable-like 'tendo Achillis' must have acted on so unusually produced a calcaneal **lever.**

The os naviculare (Plates XXV. & XXVI. fig. 1, *c*) is a transversely oblong bone, compressed from before backward. Its posterior surface is occupied by the articulation for the astragalus, which is equally divided into an inner concavity and an outer convexity, the latter approaching the conical form. From the lower part of this is continued at a right angle **a small flat** triangular surface for the cuboides (Plate XXV. fig. 3, *cb''*). The upper non-articular surface is **narrowest** at its middle, and is developed into a low oblong tuberosity on each side. On the fore part of the bone (ib. fig. 3) is the articular surface, divided into the narrow oblong tract (cm), with two slight convexities, for the mesocuneiforme (ib. fig. 1, f), and into the almost flat triangular tract (fig. 3, ci) for the ectocuneiforme (fig. 1, e). The latter surface (fig. 3, ci) is divided by a very narrow non-articular tract from the cuboidal surface (fig. 3, cb''). The inner (tibial) part of the anterior surface is non-articular, and extends beyond the mesocuneiform articulation to where the entocuneiforme would have been, had that bone existed in the tarsus of the Megatherium.

The mesocuneiforme* (Plates XXV. & XXVI. fig. 1, f) is a laterally compressed reniform bone, with a thick rough convex inferior border: the inner (tibial) surface (Plate XXV fig. 1, f) is irregular and flattened: the similarly modified outer (fibular) surface (Plate XXV. fig. 2, f) is varied by a flat elliptic articular facet (iii) near its upper part, for a similar surface on the side of the metatarsal of the 'digitus medius.' The back part of the bone is chiefly occupied by the narrow surface for the naviculare. The fore part of the bone is obtuse and rough. Not a vestige of the toe (digitus secundus), usually supported by the mesocuneiforme, is developed in the Megatherium.

The ectocuneiforme† (Plates XXV & XXVI. fig. 1, e) presents the normal wedge-like figure of the tarsal cunciform bones. It is flattened from before backward; with its thick base (e) rough and convex. The posterior surface presents the almost flat, slightly concave articular surface for the naviculare. The outer border or surface is rough and tuberous; the inner one less rough and flat. The anterior convex facet (Plate XXV. fig. 2, iii) for the base of the metacarpal does not extend to the upper or under borders of the bone. The articulation with the contiguous cuneiforme, as with the cuboides, is by syndesmosis.

The cuboides (Plate XXVI. fig. 1, d) presents an articular surface divided into three facets on its upper, or tibial, and back part; the anterior and smallest facet is for the naviculare (c), the middle and largest for the astragalus (a), and the posterior for the calcaneum (tb): the latter surface is at almost a right angle with the astragalar one, and looks backward. On the under, or fibular, and fore part of the bone is the articular surface for the two outer metatarsals; that for the fourth toe (Plate XXV. fig. 2, iv) being concave transversely and slightly convex lengthwise; that for the fifth toe (ib. v) being uniformly but very slightly convex.

A broad non-articular surface, rough and with two oblique low ridges on the upper and outer part of the bone, divides the back from the front articular surface; a narrower non-articular tract, but produced into a strong obtuse ridge, divides the same surfaces on the inner or under side of the bone. The fore part of the bone is produced into an angular process (Plate XXV. fig. 2, p), which forms the inner part of the articular channel for the fourth metatarsal. The under part of the bone is impressed by the broad tendinal groove continued from that which impresses the outer part of the calcaneum.

* 'Os cuneiforme medium' of Sœmmerring.   † 'Os cuneiforme externum,' ib.

The above-described composition of the tarsus of the Megatherium has been deduced from the study of three entire specimens of the bones of the hind foot. The distal articular surfaces demonstrate that the digits of that foot were but three in number, and that they answered to the 'third,' 'fourth,' and 'fifth' of the pentadactyle type. Not a rudiment of the 'second' exists independently, and every vestige of the first, together with the cuneiform bone supporting it, is absent. There are no little bones missing on the inner side of the 'mesocuneiforme,' as Dr. PANDER conjectured might be the case in the Madrid skeleton; and there is no 'os cuneiforme' for the hallux ('grand doit du pied'), as CUVIER supposed.

Metatarsus.—The metatarsal of the 'third' toe (Plates XXV. & XXVI. fig. 1, m_3) resembles rather an 'os cuneiforme,' by reason of its extreme shortness, or fore-and-aft compression. Its upper non-articular surface is the broadest, and is rough and convex, like that of the ectocuneiforme. The proximal or posterior surface for that cuneiform bone is triangular and slightly concave (Plate XXVI. fig. 3, *ci*); the bone is prolonged into a tuberous process beneath it, forming a lever of advantage for the insertion of a flexor tendon. On the inner side of the bone, at its upper part, is the small flat surface (*cm*) adapted to that on the mesocuneiforme; on the outer side of the bone is a larger surface, partly convex, partly concave, for articulating with the side of the base of the fourth metacarpal (Plate XXVI. fig. 1, m_4); the rest of the outer surface is very irregular, as if honeycombed. The distal or anterior articular surface presents a vertical median prominence, passing into a partly flat surface internally, and into a vertically concave surface externally. The lower part of the vertical prominence is most produced (*ib.* fig. 2, *p*).

The fourth metatarsal (Plate XXVI. fig. 1, m_4) has its base compressed laterally, and expanded vertically; but the distal end is still more produced in that direction, so as to be almost hammer-shaped. The vertically extended articular surface at the base of this metatarsal is narrow, convex transversely, and adapted to the channel in the cuboides (Plate XXV. fig. 2, *p iv*). On the inner or tibial side of the base is the articular surface, concave posteriorly, less convex anteriorly, for the interlocking joint with the 'third' metatarsal. In front and below this articular surface is a very rough honeycombed tract of bone for firm syndesmotic junction with the similarly modified surface of the third metatarsal.

On the outer or fibular side of the base are two articular surfaces for the fifth metatarsal: the hinder and smaller one is flat, and is continued, at right angles, with the basal surface; the larger surface is subcircular and rather undulating: in front and below these surfaces is a narrow rugged tract for ligamentous junction with the fifth metatarsal. The part of the shaft of the fourth metatarsal which stands out free is smooth upon its inner and upper sides, is traversed by a wide oblique tendinal groove below, and is rather rough and irregular externally.

The vertically produced distal surface presents a large rough protuberance at its upper part (Plate XXV. fig. 1, *iv*), and three protuberances at its under part, of which the

L

innermost is the most produced, the outermost the least. The articular surface resembles in character that on the end of the third metatarsal, but is relatively smaller; it consists of a longitudinal median prominence, continuous with a smooth narrow tract on the inner side, on which surface articular cartilage and synovial membrane seem to have existed at only a small part at its lower end. On the outer side, the smooth tract is limited to the upper half of the prominence; this is slightly concave vertically or lengthwise. It is a form of articulation calculated rather for firm and unyielding junction than for flexibility.

The outermost metatarsal (Plate XXVI. fig. 1, m 5), answering to the fifth of the pentadactyle foot, is the longest, but from its more backward articulation with the tarsus it does not reach so far forward as the fourth.

The proximal end is a free rough tuberosity, somewhat more than an inch anterior to which, on the tibial side, is the oval slightly concave articular surface adapted to the cuboides (Plate XXV. fig. 2, v), which surface is continuous with the short transverse flattened surface for the fourth metatarsal; and, in advance of this, is the larger, subcircular, slightly undulated surface for the corresponding surface on the fourth metatarsal. The second surface is on the middle of the inner surface of the fifth metatarsal, and is bordered in front and beneath by the very rough honeycombed surface for ligamentous junction with the adjoining metatarsal. This combination of synovial with syndesmotic joints admits of that degree of slight elastic movement of the very firmly attached fourth and fifth metatarsals, which must have facilitated the heavy tread of the ponderous quadruped, and have alleviated the effect of the enormous pressure on the two gradatorial toes. The whole outer and under part of the non-articular surface of the fifth metatarsal is strongly sculptured by irregular ridges, tubercles, grooves, and foramina, indicative of the hoof-like callosity of the outer border of the sole in which it was chiefly imbedded. The upper and inner non-articular surfaces are comparatively smooth. The inner side is flat, and is traversed by an oblique shallow (tendinal ?) groove. On the distal surface are two small articular facets; the inner one, which is the best marked, is subcircular, about 8 lines in diameter.

Phalanges.—The bones of this class are unusually reduced in number in the foot of the Megatherium, even admitting the accuracy of the figures of the hind foot of the Madrid skeleton, in which two stunted phalanges appear to terminate both the fourth and fifth toes. For, as the great unguiculate toe, like that of the fore-foot, has only two moveable phalanges, the total number of these bones is but six, not exceeding that of the tarsal bones of the same foot.

The phalanx (Plate XXVI. figs. 1 & 3, i & 3) of the innermost toe (*iii*), answering to the third of the pentadactyle foot, represents, as in the corresponding digit of the fore foot, the proximal and middle phalanges connate. The compound bone (*ib.* fig. 3, i & 3) is shaped like a wedge with an oblique edge, deeply notched. It has articular surfaces not only on its proximal and distal ends, but upon its upper or dorsal surface. The two former surfaces, which are the 'ends' of the bone in ordinary phalanges, here form the 'sides' of the wedge: the sides of the phalanx are the 'margins' of the wedge: the dorsal surface forms the base, the plantar or under surface is represented by the two processes of the cleft apex or edge of the wedge. The proximal articular surface presents a longitudinal channel, convex vertically, concave transversely, from which a flat surface extends from nearly the whole of the tibial side, and a convex surface from the upper part of the fibular side; the whole articular surface being the counterpart of that on the metatarsal (ib. *m*), with which the present remarkable bone is, by this interlocking joint, firmly united. The lower prominence of the median rising of the distal joint of the metatarsal (*p*) protrudes through the lower notch in the phalanx (*n*). A yielding, elastic, slightly-sliding movement was all that could take place between these bones. The complex distal articulation is adapted to an equally restricted junction with the enormous terminal phalanx (3); it consists of four distinct articular surfaces, two on the anterior and two on the upper part of the connate phalanges (1 & 2). The internal of the two distal surfaces is the largest and is slightly concave, the external one, near the upper and outer angle of the bone, is slightly convex; they are divided by a rough tract, indicating strong ligamentous union with the claw-phalanx, of half an inch in breadth; and they cover the smaller proportion of the distal surface of the bone. Below the outer articular surface there is a protuberance, with a smooth but non-synovial surface in the present bone, which is adapted to a definite smooth surface upon the claw-phalanx. The two upper surfaces are subcircular, each about half an inch in diameter, very slightly convex, and about 5 lines apart.

The great terminal phalanx of the present toe (Plates XXV. & XXVI. *iii*, 3) is modified, like its homotype on the fore foot, for the firm fixation and development of a powerful claw, the hollow base of which covered the bony core, and was encompassed by a bony sheath of 5 inches in length.

The base of the phalanx, whence both core and sheath extend forward, is a surface of a long, narrow, vertically elliptical form, with the upper third produced backward so as to overhang or cover the confluent supporting phalanges; and the lower third sloping forward at nearly the same angle with the vertical middle third, but terminated by a backwardly projecting ridge. There are two articular surfaces on the middle division, two on the upper division, and one on the lower division of the base. The innermost of the middle articulations is slightly convex, the outermost as slightly concave; they both look directly backward. The lower articular surface is flat and subcircular; it is on the hinder and outer angle of the lower third of the base, and looks downward and a little backward; the two small surfaces on the upper third of the base are subcircular and flat, and look downward: these different articular surfaces are counterparts of those on the distal and upper parts of the connate phalanges. On each side of the lower third of the base of the claw-phalanx is a large canal, leading forward to the interspace between the bony core and sheath of the claw, and giving passage to the vessels and nerves of the formative matrix of that instrument. The tibial side of the sheath (Plate XXV. fig. 1, 3) is gently convex; the fibular side (Plate XXVI. fig. 3, 3) is flatter;

a similar modification affects the claw-core, the narrow basal part of which is divided by sharp borders from the sides. The point of the core projects about 3 inches in advance of the lower border of the sheath: its form, broken in the specimen, is restored in dotted outline in the figures.

The distal articular surface of the fourth metatarsal shows that it supported a phalanx so articulated with it as to have no movements of flexion or extension, and only a slight degree of bending from side to side. The illustrations of the Madrid skeleton, especially plate 5, fig. 5, show that this phalanx (Plate XXVI. fig. 1, iv, 1) was very short, and that it supported a second phalanx (ib. iv, 2) of a subhemispheric form, terminated obtusely.

The distal articular surface of the fifth metatarsal indicates that it supported a phalanx smaller than the proximal one of the fourth toe; and the figure, above cited, of the Madrid skeleton shows such a phalanx (ib. v, 1), and also a second small stunted hemispheric phalanx (ib. v, 2)*; and this, from the analogy of the Mylodon, is most probably the true structure.

If we contemplate the bones of this singularly constructed foot in their natural co-adaptation, the same relation of the osseous masonry to the transference of pressure from the leg to the outer border of the foot will be appreciated, as has been pointed out in the fore limb. The broad surface of the astragalus—the great keystone of the tarso-metatarsal arch (Plate XXVI. fig. 1, a)—transmits the superincumbent weight in two chief directions, backward upon the massive heel-bone (b), forward upon the metatarsus. By the naviculare (c) it is transmitted through the ectocuneiforme (e) and the produced outer angle of the base of the mid-metatarsal (m 3) to the fourth (m 4), and thence to the fifth metatarsal (m 5). The cuboides (d), receiving the weight from both astragalus and naviculare, transmits it by its produced fore part to the base of the fourth metatarsal; and partly by that medium, but chiefly by direct articulation, to the side of the base of the fifth metatarsal. The tendency in the cuboides to yield under this pressure and slip back, is resisted by the abutment of the calcaneum (b, t) against its back part.

§ 9. *Comparison of the Bones of the Hind Foot.*

No known recent Mammal offers in its astragalus any repetition of the peculiarities of that bone in the Megatherium. In the Anteaters and Armadillos the upper surface of the astragalus has the usual configuration, and is received into a deep tibio-peroneal mortice. In the Sloths the outer part of the astragalus is excavated by a deep cell, in which the pivot-shaped end of the fibula rotates: the inner side is applied, as usual,

---

* In the original memoir, by Bru and Garriga, on the Skeleton of the Megatherium at Madrid, the metatarsus and toes of the hind foot are said to agree with those of the fore foot, except that there is only one toe with a claw, instead of three ("Tambien se advierte que en esto hay solo un *Dedo* con uña, quando en la *Mano* se registran tres; en lo demas convienen en un todo," p. 16). In the description of the outermost (fourth developed) toe of the fore foot it *is* stated that it has two phalanges, and that there is nothing to be remarked except that they are rounded ("Los dos *Falanges* del *quarto* que se reconocen en J. y L. no tienen cosa que advertir mas que son casi redondas," p. 13).

against a malleolar process of the tibia. There is an analogy to the megatherian structure in the pivoted part of the articulation, but the process and the cavity are on reverse parts of the ankle-joint.

The convex protuberance or pivot on the inner half of the tibial surface of the astragalus is common to all the extinct Megatherioids hitherto discovered, but it is associated with well-marked modifications of the bone in each genus, and with minor differences in different species. In the Mylodon * the calcaneal surface is single, and is continuous with the navicular one; no part is insulated by a bisecting groove, as in the Megatherium: the middle division (a) of the upper articular surface is less convex; the upper half of the navicular surface is flat, instead of being concave.

The astragalus of the Scelidotherium† agrees with that of the Mylodon in the less depth of the middle division of the upper surface, and the more open angle at which it joins the inner convexity; it agrees with that of the Megatherium in the division of the calcaneal surface; but it differs from both in the presence of two deep concavities upon the naviculo-cuboid surface, the portion to which the cuboides articulates being concave instead of convex.

As the modification of the calcaneal surface of the astragalus governs that of the co-adapted surface in the calcaneum, this bone, in the Mylodon, is distinguished by the uninterrupted continuity of the **articulation presented to the astragalus**, with which that for the cuboides is continuous; the posterior part of the astragalar surface is less convex than in the Megatherium: the posterior prolongation of the calcaneum is relatively shorter and is less pointed: the posterior wall of the great outer tendinal groove is more developed in the Mylodon. In the Scelidotherium the posterior termination of the calcaneum is broader, and terminated by a less angular convexity than in the Mylodon; but in the separation of the posterior from the anterior part of the astragalar surface it agrees with the Megatherium and with the Sloths. These, of all recent *Bruta*, most resemble the Megatherioids in the posterior prolongation of the calcaneum, and more especially the Megalonyx in the compression and distal expansion of that lever; but, in the Megalonyx, this expansion in the vertical direction is extreme, and gives the heel-bone more the form of an os ilium than of a tarsal bone.

In the os cuboides of the Mylodon the two surfaces for the fourth and fifth metatarsals are nearly on the same plane; in the Megatherium they are nearly at a right angle. In the Scelidotherium the surface by which the cuboides joins the astragalus is convex, instead of being concave, as it is in the Megatherium and Mylodon. In the latter genus the surface on the os naviculare for the mesocuneiform **bone** is relatively broader than in the Megatherium; and the upper division of the astragalar surface is flat, instead of being convex.

With respect to the digits, it is significant of the true affinities of the Megatherium

---

* "Description of the Skeleton of the *Mylodon robustus*," 4to, 1842, p. 117, pls. 21, 22, & 23. figs. 1 & 2.

† "Fossil Mammalia," Voyage of the Beagle, 4to, pl. 26.

to find that the Sloths alone amongst the existing BRUTA show a suppression of certain toes: in all the other existing genera of the order the hind foot retains the typical number of digits.

The Armadillos, including the *Chlamyphorus*, resemble the Megatherium in the terminal confluence of the tibia and fibula; but these bones articulate in the *Dasypodidæ* in the normal way with the astragalus, and the foot is planted firmly on the ground by its flat surface. The hallux is not only present in the *Chlamyphorus*, but is longer than the fifth toe. All the five toes are armed with claws; and anomalous as the three-toed hind feet of the Sloths may appear, they have more real resemblance with the feet of the Megatherioids than have those of any of the Armadillo family. The great extinct *Glyptodon* agrees with the *Chlamyphorus* in the pentadactyle condition of the hind foot, the os naviculare having three facets for as many cuneiform bones, the innermost of which supports a well-developed hallux. This toe is abortive in existing Sloths: it is represented, in the Ai, by a small compressed bone, consisting apparently of the connate ento-cuneiform and metatarsal: it retains its distinctness in the Unau (Plate, *Bradypus didactylus*, *i*), as the similar representative of the second toe does in the Megatherium; and the meso- and ecto-cuneiform bones are also distinct; but in the Ai each cuneiform bone is confluent with its metatarsal, with each other, with the naviculare, and with the cuboides.

In the Megatherium, the mutilation of the foot has commenced on its outer side by the removal of the ungual phalanx from the fifth and fourth toes; in the Sloths the mutilation is restricted to the fifth toe, but is carried further, viz. to the removal of all the phalanges and the reduction of size of the metatarsal (Plate, *Bradypus didactylus*, *v*): nevertheless, the Sloths alone, amongst existing *Bruta*, manifest in any degree the characteristic megatherioid condition of the marginal toes, first and fifth, of the pentadactyle series: and the minor modifications of this character, differentiating the existing from the extinct phyllophagous *Bruta*, are such as intelligibly relate to the different powers and habits of creatures so different in size and mode of progression.

The mutilation of the two outer toes in the Megatherium is accompanied by modifications which adapt them to the important office of the support and progression of the body on level ground: in the scansorial Sloths, the three middle digits being equally developed for prehensile purposes, and none being needed for walking, one toe on the outer and one on the inner side of the foot is reduced to the metatarsal basis, and is concealed beneath the skin. In the Megatherium the hallux and its cuneiform bone are wanting: the second toe is represented by its cuneiform bone, with, perhaps, a connate rudiment of a metatarsal: only the third toe can be compared, by its size and the claw it supports, with the condition of the three unguiculate toes in the Sloths; and, like those of the Ai, it has but two moveable phalanges. The reduction of the claws to one in the hind foot of the Megatherium, its enormous size and strength, its secure attachment to the phalanx, and the solidity of the articulations of the constituent bones of the whole toe, relate to the contrasted mode of obtaining the leafy food in the colossal extinct Sloth as compared with the diminutive climbing species which still exist. If the hind foot were put by the Megatherium to the preliminary work of exposing and loosening the roots of the tree about to be prostrated, its efficiency for that purpose would be greater by having but one claw, than if it had two or three: for the single claw, like a pickaxe, would clear away the soil from the interstices of the root-ramifications, the more easily by not being associated with a second contiguous claw, impeding such operation by striking upon the root itself.

§ 10. *Physiological Summary.*

To sum up the results of the foregoing descriptions of the known fossilized parts of the extinct giant of the Order *Bruta*, as indicative of its habits and affinities. The teeth agree in number, kind, mode of implantation, and growth, with those of the Sloth (*Bradypus*), and their structure is a modification of that peculiar to the same genus. All the modifications of the skull relating to the act of mastication, especially the large and complex malar bones, repeat the peculiarities presented by the existing Sloths. There are the same hemispheric depressions for the hyoid bone in the Megatherium as in the Sloth. In the number of cervical vertebræ, the Megatherium, like the Two-toed Sloth, agrees with the Mammalia generally [*]. In the accessory articular surfaces afforded by the anapophyses and parapophyses of the hinder dorsal and lumbar vertebræ[†], the Megatherium resembles the Anteaters (*Myrmecophaga*[‡]); but it does not resemble the Armadillos (*Dasypus*[§]) in having long metapophyses, the peculiar development of which in those loricated *Bruta* has a direct relation to the support of their bony dermal armour. In the mesozygapophyses of the middle dorsal vertebræ[‖], the Megatherium appears to be peculiar amongst Mammalia. In the small extent of the produced and pointed symphysis pubis it resembles the Sloths; and in the junction of both ilium and ischium with the sacrum, it manifests a character common to the Edentate order; but in the expanse and massiveness of the iliac bones, it can only be compared with other extinct members of its own peculiar family of phyllophagous BRUTA. The habits of the Megatherium necessitating a strong and powerful tail, we find this resembling in its bony structure that of other BRUTA with a similar appendage, especially in the independency of the two hæmapophyses of the first caudal, a character which obtains in the Anteaters[¶] and in some Armadillos; but this is no evidence of direct affinity to either of those families: the habits of the small arboreal Sloths render their eminently prehensile limbs sufficient for their required movements, and the tail is wanting. Had that appendage been proportionally as large as in the Megatherium, we cannot suppose that the caudal vertebræ would have materially differed from those of other BRUTA.

In the coalescence of the anterior vertebral ribs with the bony sternal ribs, the Megatherium resembles the Sloths. This essential affinity is still more marked in the pecu-

---

[*] Plate I. and *Bradypus didactylus*, C 1-7.　　　　　　　　　[†] Plate III. figs. 4 & 5, *o* and *p*.
[‡] See Philosophical Transactions for 1851, Plate XLIX. fig. 20.　　[§] Ibid. Plate XLIX. figs. 18, 19.
[‖] Plate III. *mz*.　　　　　　　　　　　　[¶] Philosophical Transactions for 1851, Plate LIII. fig. 60.

liarities of the scapula and of the carpus. In the *Myrmecophaga jubata*, the scaphoid is distinct: in the *Manis* it coalesces with the lunare: in the *Dasypus gigas* the trapezoides is anchylosed to the second metacarpal: in the *Das. sexcinctus* it has coalesced with the trapezium. Not any of these characteristics are manifested by the Megatherium; its carpus repeats the peculiarities of that in the Sloths, viz. the reduction of the number of carpal bones to seven by the coalescence of the scaphoid with the trapezium*. The first digit (pollex), which is retained in the Anteaters and Armadillos, is obsolete in the Megatherium, as in the Sloths and Orycteropus: three digits are fully developed and armed with claws, as in the *Bradypus tridactylus*; and the fifth, though incomplete in the Megatherium, is better developed, because it was required in the ponderous terrestrial Sloth for its progression on level ground. In no existing ground-dwelling member of the BRUTA is the fifth digit deprived of its ungual phalanx, as in the Megatherium. The bones of the fore foot of that extinct animal are thus seen to be modified mainly after the type of the *Bradypodidæ*.

The long bones of all the limbs are devoid of medullary cavities, as in the Sloths. The femur lacks the ligamentum teres, as in the Sloths. The fibula is anchylosed to the tibia at both ends in the Megatherium, as in *Dasypus*; but this is not the case in the closely-allied extinct Megatherioids called Mylodon, Megalonyx, and Scelidotherium, a fact which diminishes the force of the argument which CUVIER deduced from the coalesced condition of the bones in favour of the affinities of the Megatherium to the Armadillos. The semi-inverted but firm interlocking articulation of the hind foot to the leg shows the peculiarities of that joint in the Sloths exaggerated, and departs further from its characteristics in other BRUTA. In all the existing members of the order, save the Sloths, the hind foot is pentadactyle, and four of the toes have a long claw; this is even the case with the little arboreal *Myrmecophaga didactyla*. The departure by degradation from the pentadactyle type is a peculiar characteristic of the Sloth-tribe in the order: it is carried further in the same direction in the Megatherium and other great extinct terrestrial Sloths.

Guided by the general rule that animals having the same kind of dentition have the same kind of food, I conclude that the Megatherium must have subsisted, like the Sloths, on the foliage of trees; but that the greater size and strength of the jaws and teeth, and the double-ridged grinding surface of the molars in the Megatherium, adapted it to bruise the smaller branches as well as the leaves, and thus to approximate its diet to that of the Elephants and Mastodons. The Elephant and the Giraffe are specially modified to obtain their leafy food; the one being provided with a proboscis, the other with much elongated cervical vertebræ; the entire frame of the lofty ruminant being concurrently modified to adapt it to browse on branches above the reach of its largest congeners. If the Megatherium had possessed, as CUVIER conjectured, a proboscis, it could not, judging from the suborbital foramina, have exceeded in size that of the Tapir, and must have been able to operate only upon branches brought near the mouth. Of the

* Plate, *Bradypus didactylus*, et.

use of such a proboscis in obtaining nutritious roots, on the Cuvierian hypothesis that such formed the sustenance of the Megatherium, it is not easy to speculate: the Hog's snout might be supposed to be more serviceable in obtaining those buried parts of vegetables; but no trace of the prenasal bone exists in the skull of the Megatherium. A short proboscis might be useful in rending off the branches of a tree when prostrated and within reach of the low and broad-bodied Megatherium, but this office has been provided for by the organization of the tongue, of which, both the hyoid skeleton by its strength and articulation, and the foramina for the muscular nerves by their unusual area, attest the great size and power.

As regards the limbs, the Megatherium differs from the Giraffe and Elephant in the unguiculate character of certain of its toes, in the power of rotating the bones of the fore arm, in the corresponding development of supinator and entocondyloid ridges on the humerus, and in the possession of complete clavicles. These bones are requisite to give due strength and stability to the shoulder-joint for varied actions of the fore arm, as in grasping, climbing, and burrowing. But they are not essential to scansorial or fossorial quadrupeds: the Bear and the Badger have not a trace of clavicles, and merely rudiments of these bones exist in the Rabbit and the Fox. We must seek, therefore, in the other parts of the organization of the Megatherium, for a clue to the nature of the actions by which it obtained its food. In habitual burrowers the claws can be extended in the same plane as the palm, and they are broader than they are deep. In the Megatherium the depth of the claw-phalanx exceeds its breadth, especially in the large one of the middle finger; and not any of the claws can be extended into a line with the metacarpus, but they are all more or less bent inward and downward. Thus, although they might be used for occasional acts of scratching up the soil, they are better adapted for grasping; and the whole structure of the fore foot militates against the hypothesis of PANDER and D'ALTON*, that the Megatherium was a burrowing animal.

The same structure equally shows that it was not, as Dr. LUND† supposes, a scansorial quadruped; for, in the degree in which the fore foot departs from the structure of that of the existing Sloths, it is unfitted for climbing; and the outer digit is modified, after the ungulate type, for the exclusive office of supporting the body in ordinary terrestrial progression. It may be inferred from the diminished curvature and length, and from the increased strength and the inequality of the claws, especially the disproportionately large size of that weapon of the middle digit, that the fore foot of the Megatherium was occasionally applied by the short and strong fore limb in the act of digging; but its analogy to that of the Anteaters teaches that the fossorial actions were limited to the removal of the surface-soil, in order to expose something there concealed, and not for the purpose of burrowing. Such an instrument would be equally effective in the disturbance of roots and of ants; it is, however, still better adapted for grasping than for

* Das Riesen-Faulthier, &c. fol. 1821, p. 16.
† Blik paa Brasiliens Dyreverden för sidste Jordomvæltning, af Dr. LUND, 4to. Kjöbenhavn, 1838, p. 21.

delving. But to whatever task the partially unguiculate hand of the Megatherium might have been applied, the bones of the wrist, forearm, arm, and shoulder, attest the prodigious force which would be brought to bear upon its execution. The general organization of the anterior extremity of the Megatherium is incompatible with its **being a strictly scansorial or exclusively fossorial animal, and its** teeth and jaws decidedly **negative the idea of its having fed upon insects;** the two extremes in regard to the length of the jaws are presented by the phyllophagous and myrmecophagous members of the order *Bruta*, and the Megatherium in the shortness of its face agrees with the Sloths.

Proceeding, then, to other parts of the skeleton for the solution of the question as to how the Megatherium obtained its leafy food, it may be remarked that the pelvis and hind limbs of the strictly burrowing animal, *e. g.* the Mole, are remarkably slender and feeble, and they offer no notable development in the Rabbit, the Orycterope, or other less powerful excavators. In the climbing animals, as, *e. g.*, the Sloths and Orangs, the hind legs are much shorter than the fore legs; and even in those Quadrumana in which the **prehensile tail is superadded** to the sacrum, the pelvis is not remarkable for its size or the expansion of its iliac bones. But, in the Megatherium, the extraordinary bulk and massive proportions of the pelvis and hind limbs arrest the attention of the least curious beholder, and become to the **physiologist** eminently suggestive of the peculiar powers and actions of the animal. The enormous pelvis was the centre whence muscular masses of unwonted force diverged to act upon the trunk, the tail, and the hind legs, and also by the 'latissimus dorsi' on the fore limbs. The fore foot, being adapted for scratching as well as for grasping, may have been employed in removing the earth from the roots of the tree and detaching them from the soil: but the hind foot, which, like a pickaxe, had but one strong perforating and digging pointed weapon, was more probably the instrument mainly employed in removing the earth from the ramifications of the root. **The fore** limbs, terminated each by three claws, appear to have been more especially adapted for grasping the trunk of a tree; and the forces concentrated upon them from the broad posterior basis of the body must have cooperated with them in the labour, for which they are so amply organized, of uprooting and prostrating the tree. To give due resistance and stability to the pelvis, the bones of the hind legs are extraordinarily and massively developed, and the strong and powerful tail must have concurred with the two hind legs in forming a tripod, as a firm foundation for the vast pelvis, thus providing adequate resistance to the forces acting and re-acting from and upon that great osseous centre. The large processes and capacious spinal canal indicate the strength of the muscles which surrounded the tail, and the vast mass of nervous fibre from which those muscles derived their energy. The natural co-adaptation of the articular surfaces shows that the ordinary inflection of the end of the tail was backward, as in a *cauda fulciens*, not forward, as in a *cauda prehensilis*. Dr. LUND's hypothesis, therefore, that the Megatherium was a climber and had a prehensile tail, is destroyed by the now known structure of that part.

But viewing the pelvis of the Megatherium as being the fixed centre towards which the fore legs and fore part of the body were **drawn in** the gigantic leaf-eater's efforts to uprend the tree that bore its sustenance, **the colossal** proportions of its hind extremities and tail lose all their anomaly, and **appear in just harmony** with the robust claviculate and unguiculate fore limbs with which they combined their forces in the Herculean labour.

Finally, with reference to the hypothesis of the German authors and artists* of the gradual degeneration of the ancient Megatherioids of South America into the modern Sloths, it may be admitted that the general results of the labours of the anatomist in the restoration of extinct species, viewed in relation to their existing representatives of the different continents and islands, have been such as might naturally suggest the idea that the races of animals had deteriorated in point of size. Thus the palmated Megaceros is contrasted, by its superior bulk, with the Fallow-deer, and the great Cave-bear with the actual Brown Bear of Europe. The huge Diprotodon and Nototherium afford a similar contrast with the Kangaroos and Koalas of Australia, as do the towering Dinornis and Palapteryx with the small Apteryx of New Zealand. But the comparatively diminutive aboriginal animals of South America, Australia, and New Zealand, which are the nearest allies of the gigantic extinct species respectively characteristic of such tracts of dry land, are specifically distinct, and usually by characters so well marked as to require a subgeneric division, and such as no known outward influences have been observed to produce by progressive alteration of structure. Moreover, as in England, for example, our Moles, Water-voles, Weasels, Foxes, and Badgers, are of the same species as those that coexisted with the Mammoth, Tichorhine Rhinoceros, Cave Hyæna, Bear, &c.; so likewise the remains of small Sloths and Armadillos are found associated with the Megatherium and Glyptodon in South America; the fossil remains of ordinary Kangaroos and Wombats occur together with those of gigantic herbivorous Marsupials in Australia; and there is similar evidence that the Apteryx coexisted with the Dinornis in New Zealand. I have been led, therefore, to offer the following suggestions as more applicable to, or explanatory of, the phenomena than LAMARCK's progressive hypothesis of the origin of species by transmutation†, or BUFFON's retrograde hypothesis by degradation‡.

"In proportion to the bulk of an animal is the difficulty of the contest which, as a living being, it has to maintain against the surrounding influences which are ever tending to dissolve the vital bond and subjugate the organized matter to the ordinary chemical and physical forces. Any changes, therefore, in the external circumstances in which a species may have been adapted to exist, will militate against that existence in probably a geometrical ratio to the bulk of such species. If a dry season be gradually prolonged, the large Mammal will suffer from the drought sooner than the small one; if such alteration of climate affect the quantity of vegetable food, the bulky Herbivore will first feel the effect of the stinted nourishment; if new enemies are introduced, the large and

* PANDER and D'ALTON, *loc. cit.* † Philosophie Zoologique, 8vo. 1809, tom. i. ch. vii.
‡ Histoire **Naturelle**, 4to. tom. iv. (1766), "Dégénération des Animaux," p. 311.

conspicuous quadruped or bird will fall a prey, whilst the smaller species might conceal themselves and escape. Smaller quadrupeds are usually more prolific than larger ones. The actual presence, therefore, of small species of animals in countries where the larger species of the same natural families formerly existed, is not to be ascribed to any gradual diminution of the size of such larger animals, but is the result of circumstances which may be illustrated by the fable of the 'oak and the reed;' the small animals have bent and accommodated themselves to changes under which the larger species have succumbed *."

§ 11. *Geological Summary.*

The first evidences of the great extinct quadruped which forms the subject of the present Work, were received in Europe in 1789; they consisted of a considerable proportion of the skeleton of a full-grown individual, which had been discovered " in some excavations on the banks of the river Luxan, which flows close by the town of the same name, about thirteen leagues W.S.W. of Buenos Ayres, in a ravine ten yards in depth, which lies at a league and a half to the S.W. of the said town†:" and were transmitted by His Excellency the Marquis of LORETO, Viceroy of Buenos Ayres, to the Royal Museum of Natural History at Madrid.

A considerable portion of a second skeleton of a Megatherium was transmitted from Lima to the same museum in 1795. Parts of a third skeleton were, in the same year, sent from Paraguay to the Padre FERNANDO SCIO, of the 'Escuelas Pias‡.' The particular circumstances attending the discovery of the last two specimens are not recorded. The skeleton, mounted, from the foregoing materials, forms the subject of the work by BRU and GARRIGA above cited, of the memoirs by CUVIER, and of the illustrations by PANDER and D'ALTON, cited at p. 5.

In 1823, a Prussian traveller in South America, SELLOW, discovered in the bed of the Queguay, Uruguay, part of the femur of a Megatherium: it is deposited in the Royal Museum of Natural History at Berlin, and is noticed by Professor WEISS, in his Memoir in the 'Berlin Transactions' for 1827, quoted at page 6 of the present Work.

In 1824, Messrs. MITCHELL and COOPER gave a brief notice of some bones and teeth of the Megatherium, discovered in the pleistocene marl of the Skidaway Marshes, in the State of New York, North America§. In the same year, Messrs. WARING and HABERSHAW ‖ obtained a few bones of the Megatherium from a like recent formation at 'White-Bluff,' Savannah. In 1831 an instructive proportion of the skeleton of a Megatherium, including parts not preserved in the Madrid specimen, were discovered in the bed of

---

* " On the genus *Dinornis* " (part 4), Transactions of the Zoological Society, vol. iv. p. 15

† " En las escavaciones que se hacian en las orillas del Rio Luxan, que corre immediato á la Villa de este nombre (que está á unas trece leguas O.S.O. de Buenos Ayres) en una barauca de diez varas de alto, que está á legua y media al O.S. de dicha Villa."—Descripcion del Esqueleto de un Quadrúpedo muy corpulento y raro, &c., Don JOSEPH GARRIGA, Madrid, fol. 1796, 'Prologo.'

‡ Ibid.   § Annals of the Lyceum of New York, vol. i. p. 58.   ‖ Ibid.

the river Salado, which flows through the flat alluvial plains to the south of the city of Buenos Ayres. They were placed at the disposal of Sir WOODBINE PARISH, K.H., by Don HILARIO SOSA, the owner of the property on which they were found, and were presented by Sir WOODBINE to the Royal College of Surgeons. They form the subject of the Memoir by WILLIAM CLIFT, Esq., F.R.S., published in the 'Transactions of the Geological Society,' Second Series, vol. iii. p. 437.

In 1834 Mr. CHARLES DARWIN discovered remains of the Megatherium on the shores of the bay called 'Bahia Blanca,' in latitude 39° S., about 250 miles south of the Rio Plata: they were impacted in a mass of quartz shingle, forming an irregularly stratified mass divided by curved layers of indurated clay, and constituting the lower bed of the cliff called 'Punta Alta,' about 18 miles from Monte Hermoso, North Patagonia. The pebbles are cemented together by calcareous matter, probably resulting from the decomposition of numerous imbedded shells. These remains are preserved in the Museum of the Royal College of Surgeons, and are described in the volume of the 'Fossil Mammalia of the Voyage of the Beagle,' &c., 4to. 1840, p. 100. pl. 30.

In 1837 a second nearly entire skeleton of the Megatherium was discovered by Signor MUNIZ in the river Luxan, near the locality of the first discovered specimen: this was deposited in the Museum at Monte Video. In 1838 Dr. DECALZI discovered on the estate of Don MANUEL ROSAS, called 'Las Averias,' north of the Rio Salado, various bones of the Megatherium. Other specimens, subsequently discovered at Luxan, were transmitted by Mr. FALCONET, in 1845, to London, and were purchased by the Trustees of the British Museum: they are noticed at p. 11, and form, with the specimens previously transmitted to London, the subjects of the descriptions and figures in the present work.

From the foregoing summary it appears that the Megatherium ranged from the State of New York to Northern Patagonia, being restricted by the nature of its food, its habits and organization, to the latitudes most favourable to the growth of trees.

The geological and palæontological evidences concur in showing that the formations in which its remains have been found are of the newer pliocene period, coeval apparently with those which contain the remains of the Mammoth and Megaceros in Europe. Of twenty-three kinds of shells determined by Mr. SOWERBY in the collection brought by Mr. DARWIN from the Megatherium-shingle at Punta Alta, twelve are certainly identical with existing species, and four are probably so, the doubt partly arising from the imperfect condition of the specimens. These species, moreover, still inhabit the bay on the shores of which they are found fossil.

The conclusion to which Mr. DARWIN arrived as to the comparatively recent period of the shingle strata at Punta Alta, is nearly the same as that to which he was led by observations on the beds in which the megatherian remains were imbedded in the Pampas and river-courses of the province of Buenos Ayres. "We may suppose," he writes, "that whilst the ancient rivers of the Plata occasionally carried down the carcasses of animals existing in that country, and deposited them in the mud of the estuary, other animals inhabited the plains round the Sierra de la Ventana, and that lesser streams acting together with the currents of a large bay, drifted their remains towards a point where sand and shingle were accumulating in a shoal. The whole area has since been elevated; the estuary mud of former rivers has been converted into wide and level plains; and the shoals of the ancient Bahia Blanca now form low headlands on the **present** coast... everything indicates a former state of tranquillity, during which various deposits were accumulating near the then existing coasts, in the same manner as we **may** suppose others are at this day in progress. The only physical change, which we know has taken place, since the existence of these ancient Mammalia, has been a small **and** gradual rising of the continent*."

* Fossil Mammalia of the Voyage of the Beagle, p. 11.

PLATE I.

Skeleton of the Megatherium, on the scale of one inch to a foot. (The vertebræ concealed by the scapula are added in outline.)

- C 1–7. Cervical vertebræ.
- D 1–16. Dorsal vertebræ.
- L 1–3. Lumbar vertebræ.
- S. Sacrum.
- Cd 1–13. Caudal vertebræ.
- 10. Stylohyal.
- 51. Scapula.
- 52. Humerus.
- 54. Ulna.
- 55. Radius.
- s. Scaphoid part ⎫ of Scaphotrape-
- t. Trapezial part ⎭ zium.
- l. Lunare.
- 64. Pubis.
- 65. Femur.
- 66. Tibia.
- 66'. Patella.
- 67. Fibula.
- 67'. Fabella.
- a. Astragalus.
- cl. Calcaneum.
- s. Scaphoides.
- c. Cuneiforme.
- p. Pisiforme.
- d. Trapezoides.
- m. Magnum.
- u. Unciforme.
- I. Rudimentary metacarpal of first digit or pollex.
- II. Second digit, or index.
- III. Third digit, or medius.
- IV. Fourth digit, or annularis.
- V. Fifth digit, or minimus.
- 62. Ilium.
- 63. Ischium.
- ce. Ectocuneiform.
- b. Cuboides.
- II. Rudimentary metatarsal of second digit.
- III. Third digit.
- IV. Fourth digit.
- V. Fifth digit. (There is no rudiment of the first or innermost toe of the hind foot in the Megatherium.)

PLATE II.

Typical vertebræ in the Megatherium. One-fourth natural size.

Fig. 1. Fifth dorsal vertebra, or natural segment of the skeleton. c. Centrum, c'. articular tubercle for head rib; n, neurapophysis, n', neurapophysial surface for neck of rib; d, diapophysis, d', diapophysial surface for tubercle of rib; z, anterior zygapophysis; m, metapophysis; ns, neural spine; pl, pleurapophysis or 'vertebral rib'; c'', head; n'', articular surface on upper part of the neck; d'', articular surface on tubercle; h, hæmapophysis or 'sternal rib,' s', s', its condyles; hs, hæmal spine, or 'sternal bone;' s'', s'', articular surfaces for hæmapophysial condyles.

Fig. 2. First caudal vertebra. n. Neural arch; p, parapophysial part, d, diapophysial part, pl, pleurapophysial part, of compound transverse process; hy, hypapophyses; h, hæmapophysis, hy', articular surface for its own vertebra, which is in advance; hy'', articular surface for succeeding vertebra.

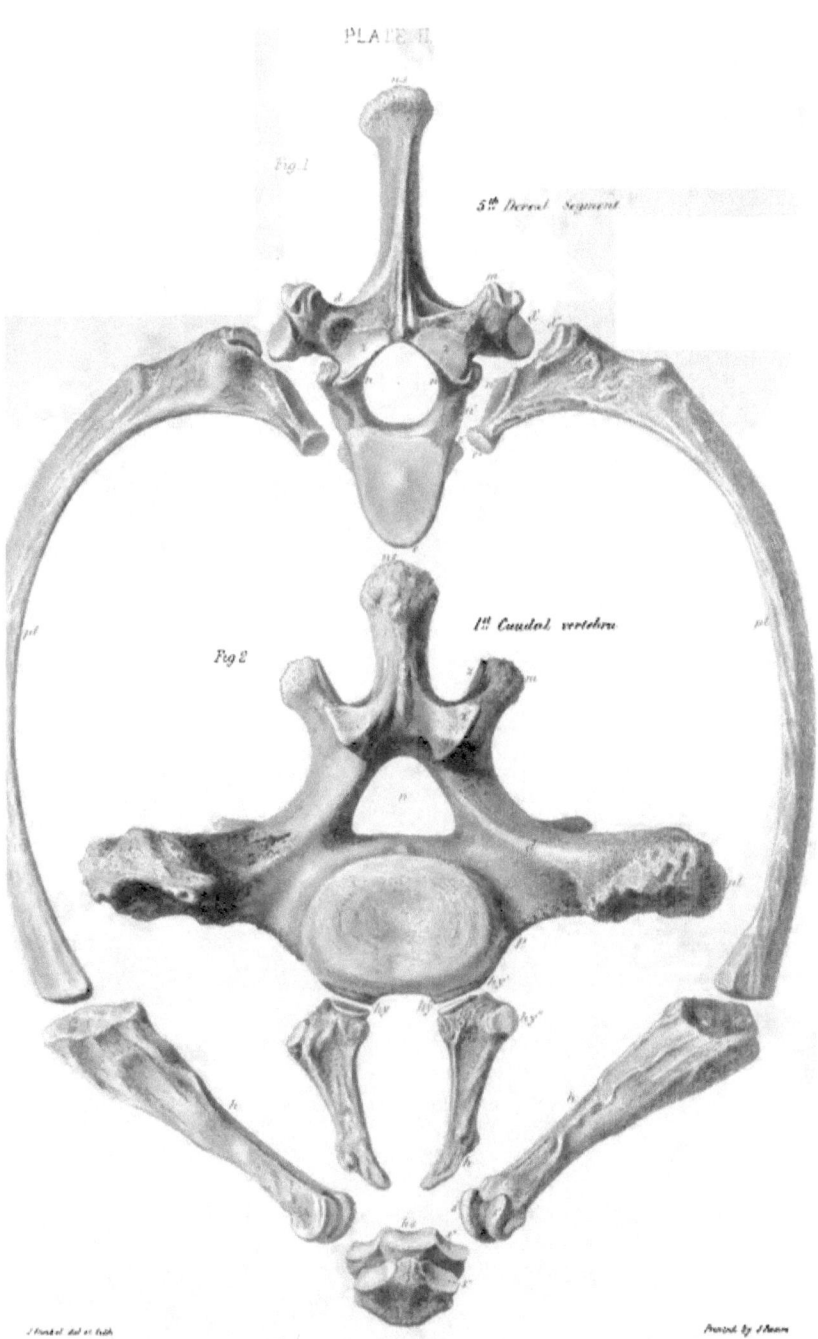

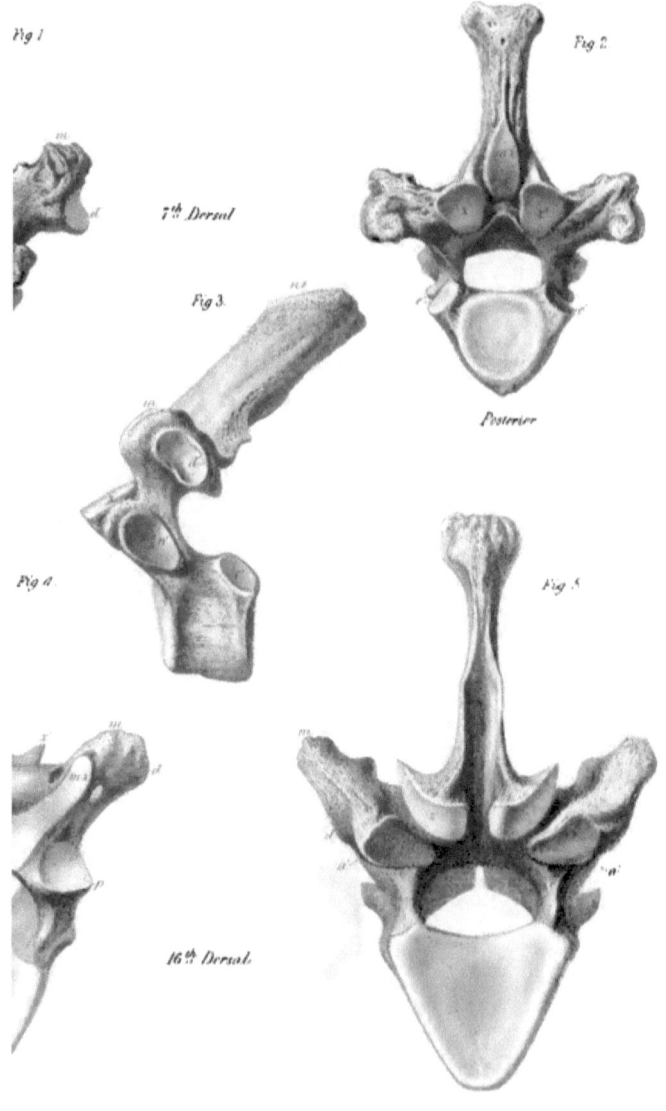

PLATE III.

Fig. 1. Seventh dorsal vertebra, front view.
Fig. 2. Seventh dorsal vertebra, back view.
Fig. 3. Seventh dorsal vertebra, side view.
Fig. 4. **Sixteenth dorsal vertebra, front view.**
Fig. 5. Sixteenth dorsal vertebra, back view.

All the figures are **drawn, minus the hæmal arch, one-fourth natural size.**

c'. Articular tubercle for the head of the rib; n', neurapophysial articular surface for the neck of the rib; d, diapophysis, d', diapophysial articular surface for tubercle of rib; m, metapophysis, mz', metapophysial articular surface; a, anapophysis, a', anapophysial articular surface; p, parapophysis, pa, parapophysial articular surface; z, anterior, z', posterior, mz, mid-anterior, mz', **mid-posterior**, zygapophyses; ns, **neural spine.**

PLATE IV.

Fig. 1. The atlas, upper view.
Fig. 2. The atlas, under view.
Fig. 3. The atlas, back view.
Fig. 4. The atlas, side view.
Fig. 5. The seven cervical and first dorsal vertebræ, upper view.
Fig. 6. The seventh cervical vertebra, back view.
Fig. 7. The seventh cervical vertebra, side view.

One-fourth natural size; c, centrum; n, neurapophysis; ns, neural spine; d, diapophysis, d_1–d_7, of the third to the seventh cervicals inclusive; pl, pleurapophysis; pl', (fig. 7) articular surface for head of first dorsal rib; m, metapophysis; nz (fig. 2), anterior articular surface; z', posterior zygapophysis; hy, hypapophysis; o, (figs. 1 & 3) its articular surface for the odontoid or true body of the atlas; r, division of the foramen alare posterius, for the passage of the posterior branch of the occipital artery; s, division of the same foramen for the passage of a branch of the second spinal nerve; v, foramen alare anterius, communicating with q the canal for the vertebral artery.

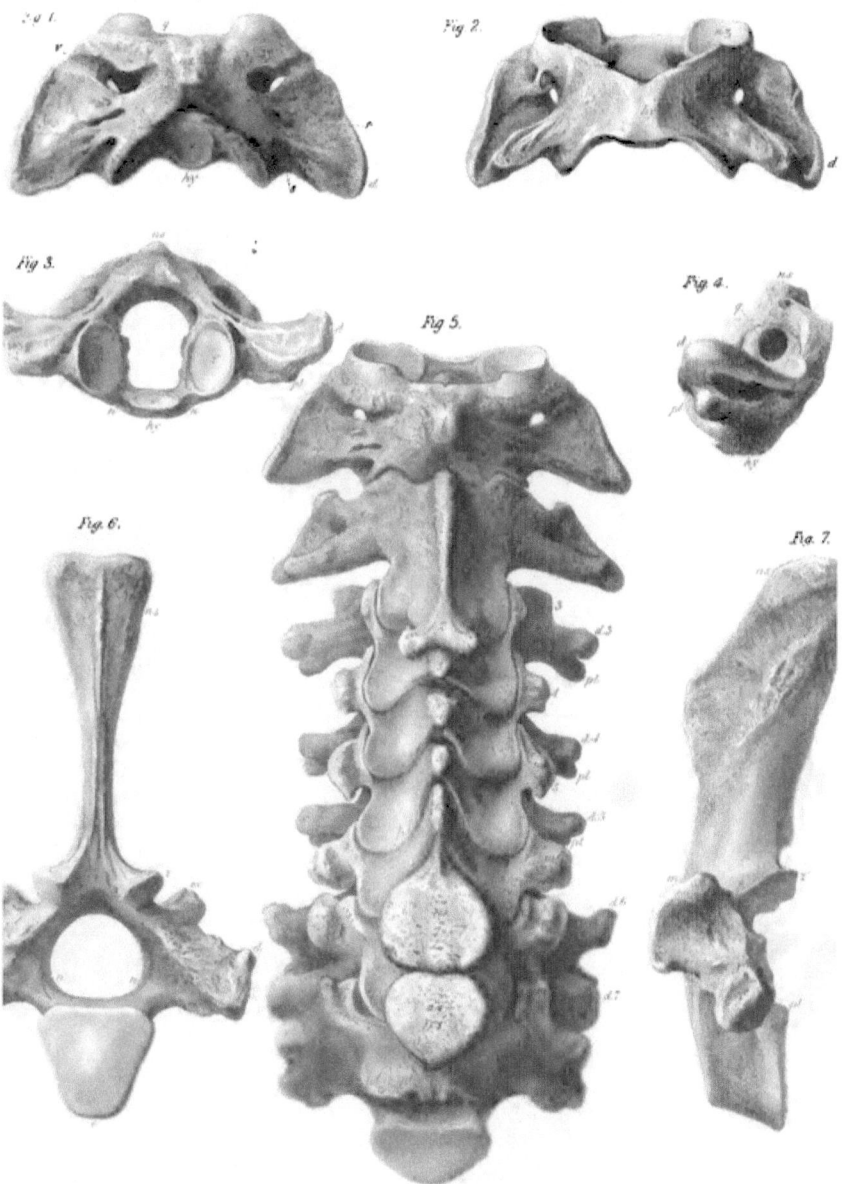

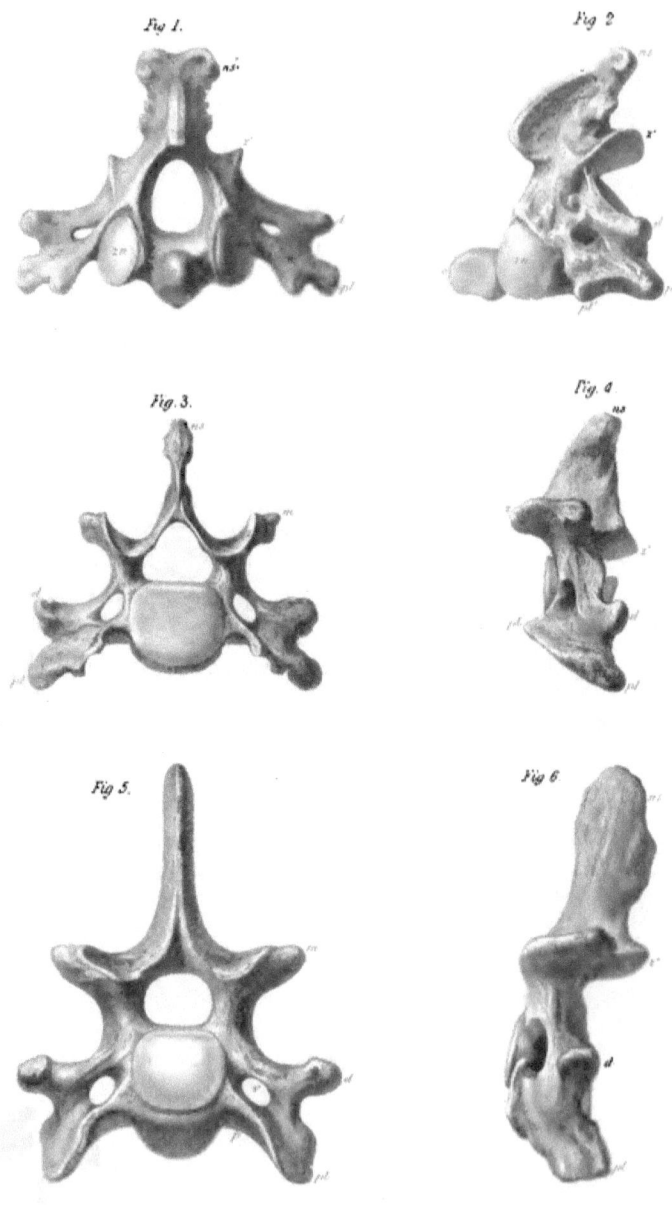

PLATE V.

Fig. 1. The axis, or vertebra dentata, front view.
Fig. 2. The axis, or vertebra dentata, side view.
Fig. 3. The third cervical vertebra, front view.
Fig. 4. The third cervical vertebra, side view.
Fig. 5. The sixth cervical vertebra, front view.
Fig. 6. The sixth cervical vertebra, side view.

    One-fourth natural size; o, odontoid process (centrum of atlas); za, analogues of anterior zygapophyses, in the dentata; z, anterior zygapophysis; z', posterior zygapophysis; d, diapophysis; p, parapophysis; pl, pleurapophysis, pl', its anterior production; m, metapophysis; ns, neural spine.

PLATE VI.

First sacral vertebra, with part of its hæmal arch (ilium and pubis).

One-fourth natural size; *c*, centrum; ** exogenous growths; *ma*, metapophysial process and articular surface.

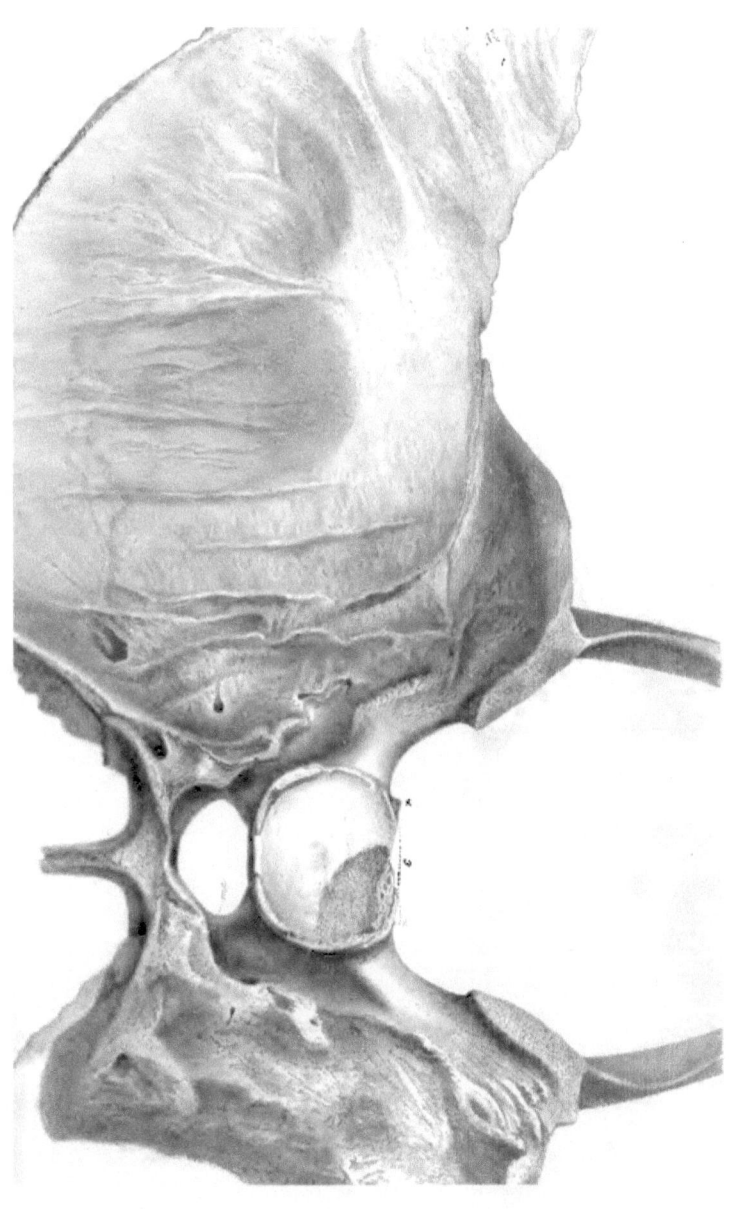

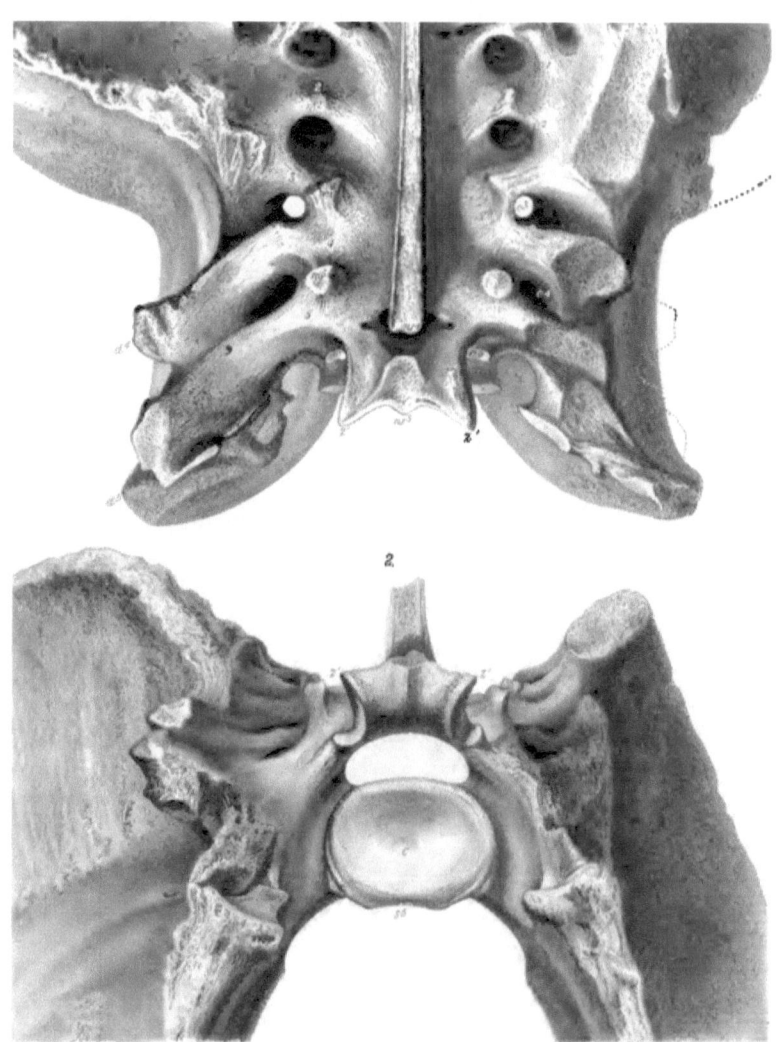

PLATE VII.

Fig. 1. The five sacral vertebræ, upper view.
Fig. 2. The sacral vertebræ, back view.

One-fourth natural size; c, centrum; n_1, n_4, coalesced neurapophyses; ns_1, ns_4, coalesced neural spines, of the four anterior vertebræ; ns_5, neural spine of fifth vertebra; $d_1, _2, _3, _4, _5$, diapophyses; m_1-m_5, metapophyses; o_1-o_5, posterior outlets of nerve-canals; z, anterior, z', posterior, zygapophyses.

PLATE VIII.

Fig. 1. The second caudal vertebra, back view.
Fig. 2. The second caudal vertebra, under view, minus the hæmal arch.
Fig. 3. The eleventh caudal vertebra, back view.
Fig. 4. The eleventh caudal vertebra, side view.
Fig. 5. The eleventh caudal vertebra, upper view.
Fig. 6. The eleventh caudal vertebra, under view, minus the hæmal arch.
Fig. 7. The thirteenth caudal vertebra, upper view.
Fig. 8. The sixteenth caudal vertebra, upper view.
Fig. 9. The sixteenth caudal vertebra, under view.
Fig. 10. The seventeenth caudal vertebra, upper view.
Fig. 11. The seventeenth caudal vertebra, under view.
Fig. 12. The eighteenth caudal vertebra, back view.
Fig. 13. The eighteenth caudal vertebra, front view.

One-fourth natural size; c, centrum; hy, anterior, hy', posterior hypapophyses; h', their hæmal articular surface; n, neural arch and neurapophyses; ns, neural spine; d, diapophysis; pl, pleurapophysis; m, metapophysis; z, anterior, z', posterior zygapophysis; h, hæmal arch, hy', (fig. 1) its anterior hypapophysial articular surface, hy'', its posterior hypapophysial surface; hs, hæmal spine.

PLATE. VIII.

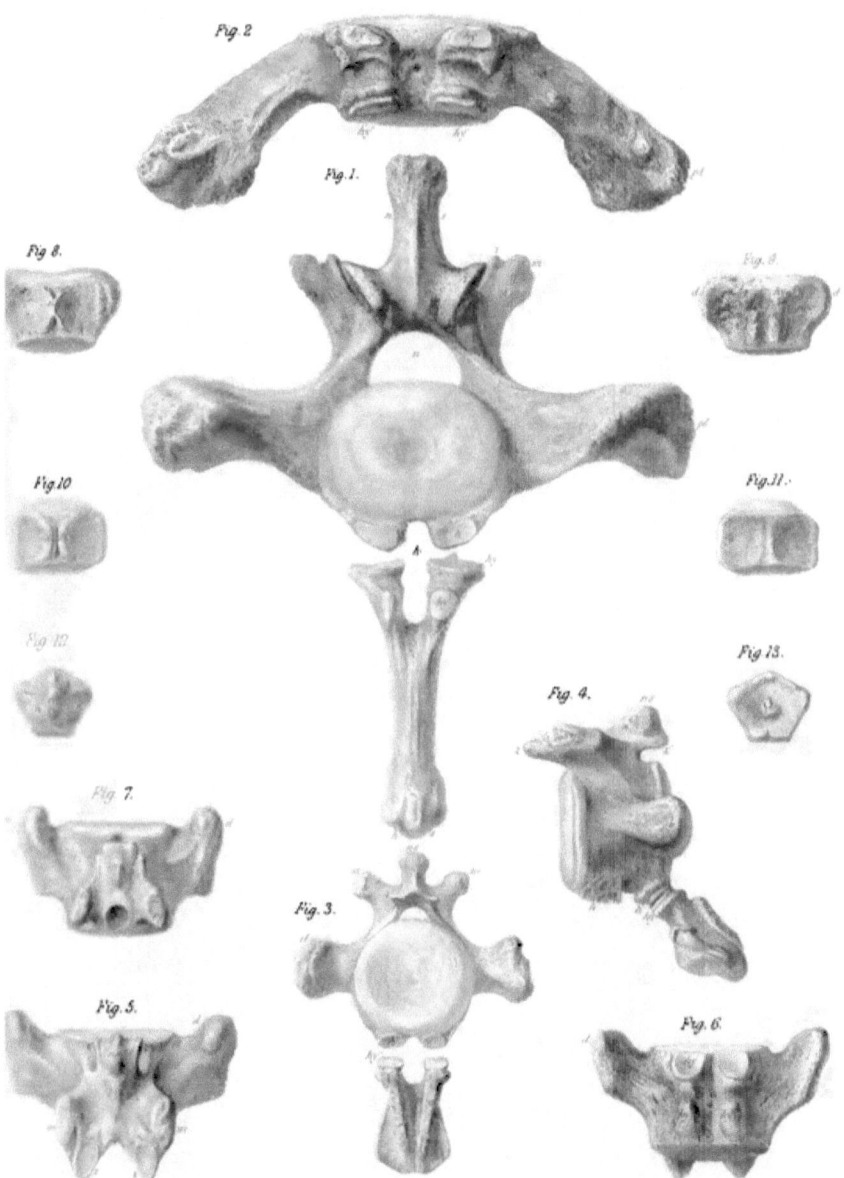

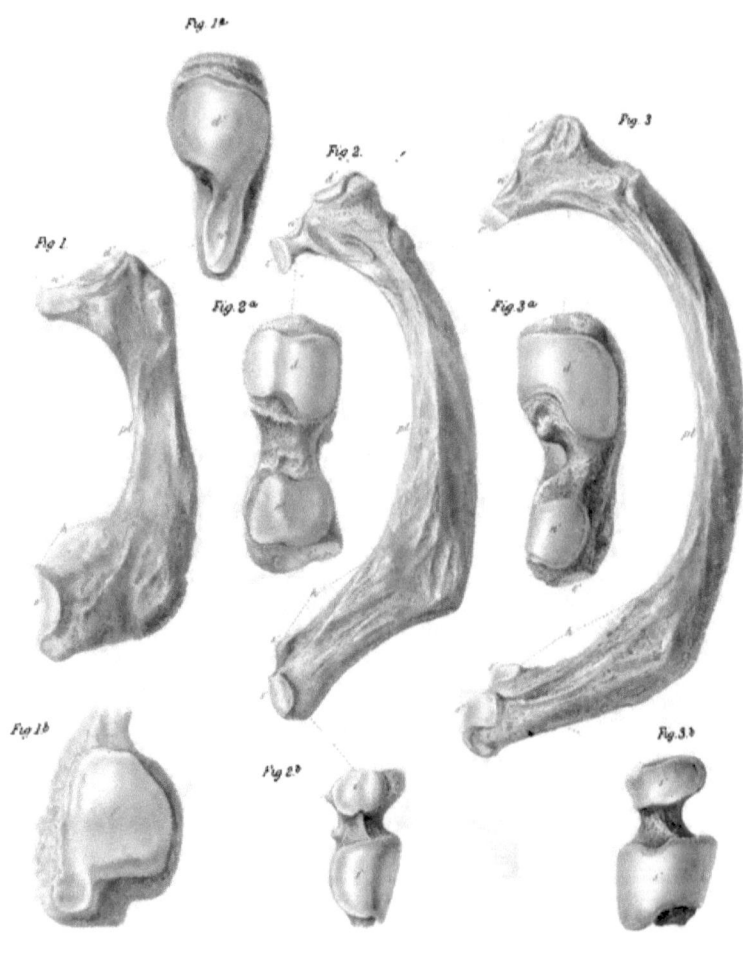

PLATE IX.

Fig. 1. First dorsal rib.
Fig. 2. Second dorsal rib.
Fig. 3. Third dorsal rib; one-fourth natural size.
Figs. 1a, 2a, 3a, their upper articulations.
Figs. 1b, 2b, 3b, their lower articulations; one-half natural size.

c'', articular surface on the head for the centrum; n'', articular surface on the neck for the neurapophysis; d'', articular surface on the tubercle for the diapophysis; s', anterior condyle for sternum; s'', posterior condyle for sternum; pl, pleurapophysial part, h, hæmapophysial part, of rib.

PLATE X.

Fig. 1. Head and neck of the fifteenth dorsal rib, upper view.
Fig. 2. Head and neck of the sixteenth dorsal rib, upper view.
Fig. 3. Head and neck of the fifteenth dorsal rib, side view.
　　　　One-half natural size; n'', neurapophysial articular surface.
Fig. 4. The thirteenth dorsal vertebra.
Fig. 5. The sixteenth dorsal vertebra.
Fig. 6. The first caudal vertebra, **under** view, minus the hæmal arch.
Fig. 7. The clavicle.
　　　　One-fourth natural **size**; c', articular **surface for the head of the rib**; n', neurapophysial **surface** for the **neck of the rib**; p, parapophysis; d, diapophysis; a, anapophysis, a', **its** articular surface; pl, pleurapophysis; hy, hypapophysis; z, anterior, z', posterior zygapophysis; ns, neural spine.

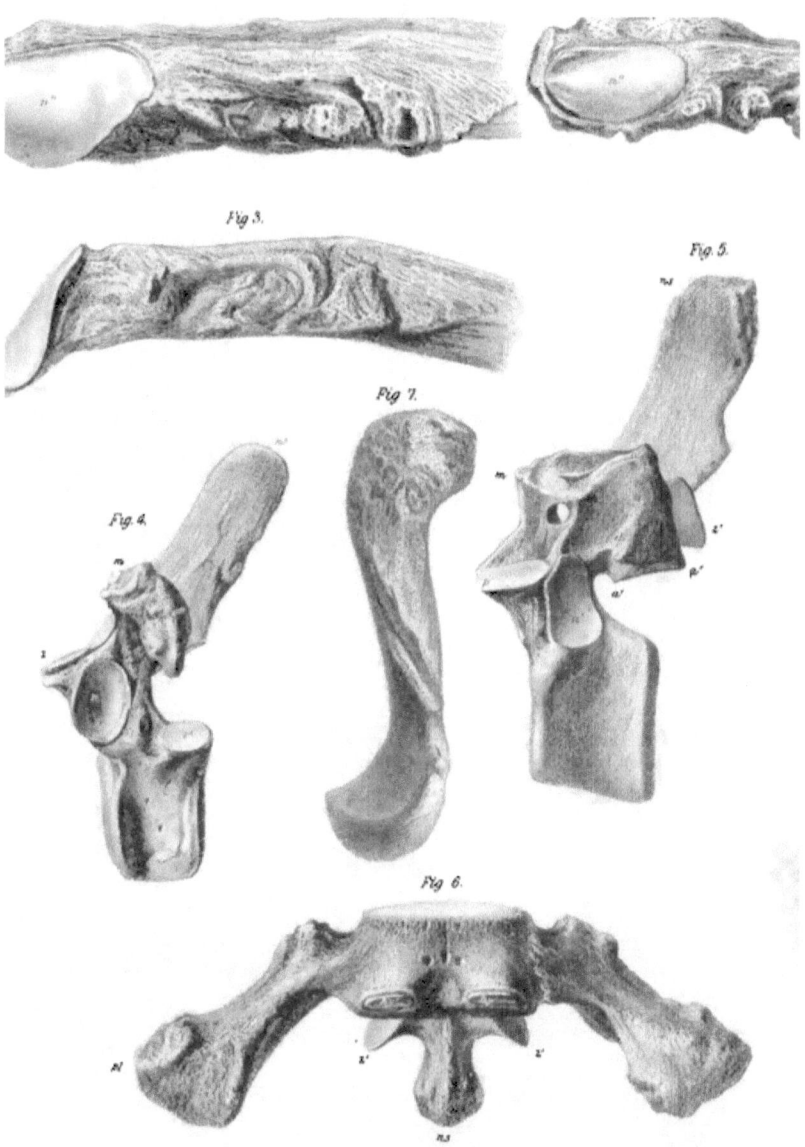

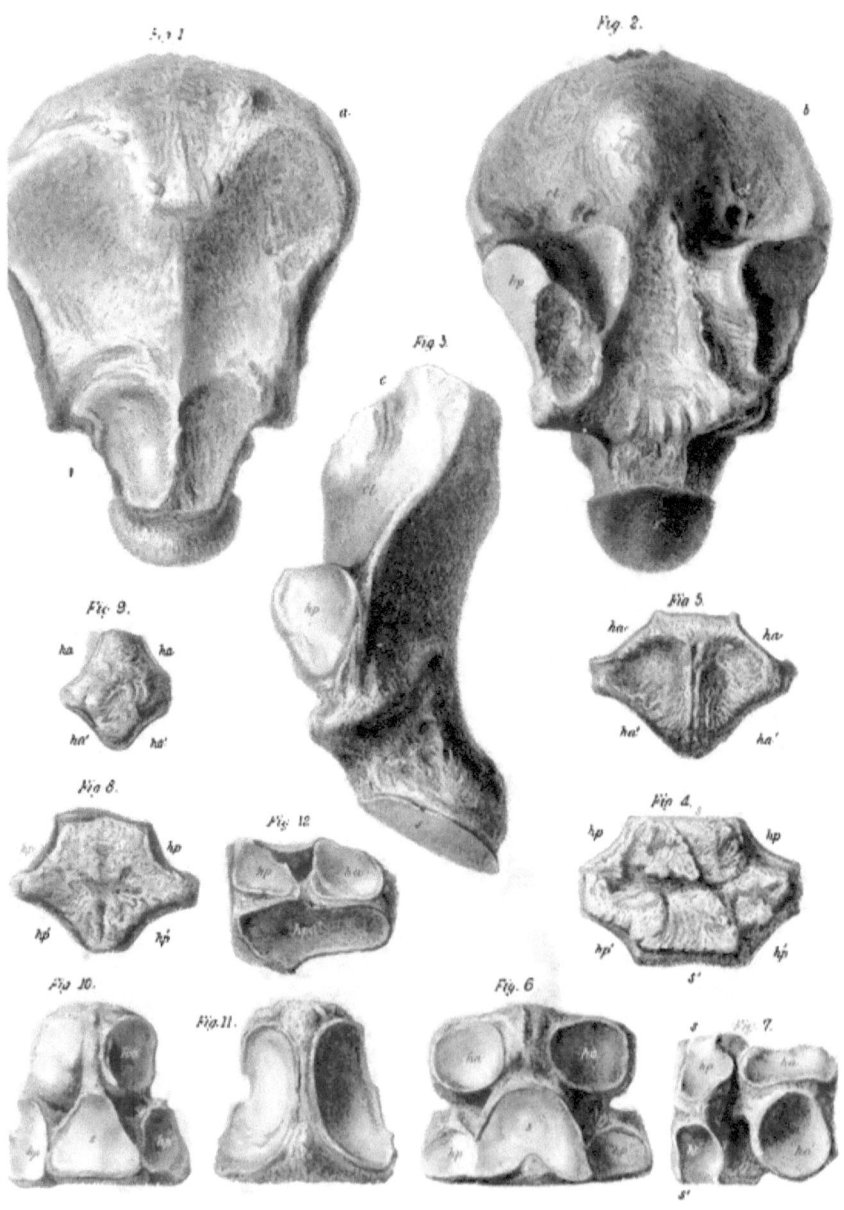

PLATE XI.

Hæmapophyses of dorsal vertebræ or bones of the sternum.

Fig. 1. Manubrium sterni, outer or under view.
Fig. 2. Manubrium sterni, inner or upper view.
Fig. 3. Manubrium sterni, side view.
Fig. 4. Sixth sternal bone, inner or central surface.
Fig. 5. Sixth sternal bone, outer or peripheral surface.
Fig. 6. Sixth sternal bone, upper or anterior surface.
Fig. 7. Sixth sternal bone, side view.
Fig. 8. Eighth sternal bone, inner surface.
Fig. 9. Eighth sternal bone, outer surface.
Fig. 10. Eighth sternal bone, upper surface.
Fig. 11. Eighth sternal bone, under surface.

One-fourth natural size; cl, surface for the clavicle; hp, articular surface for hæmapophysis, or sternal portion of rib; s and s', articular surface for contiguous sternal bone; ha, anterior outer hæmapophysial surface, ha', posterior outer hæmapophysial surface; hp, anterior inner, hp', posterior inner hæmapophysial surface.

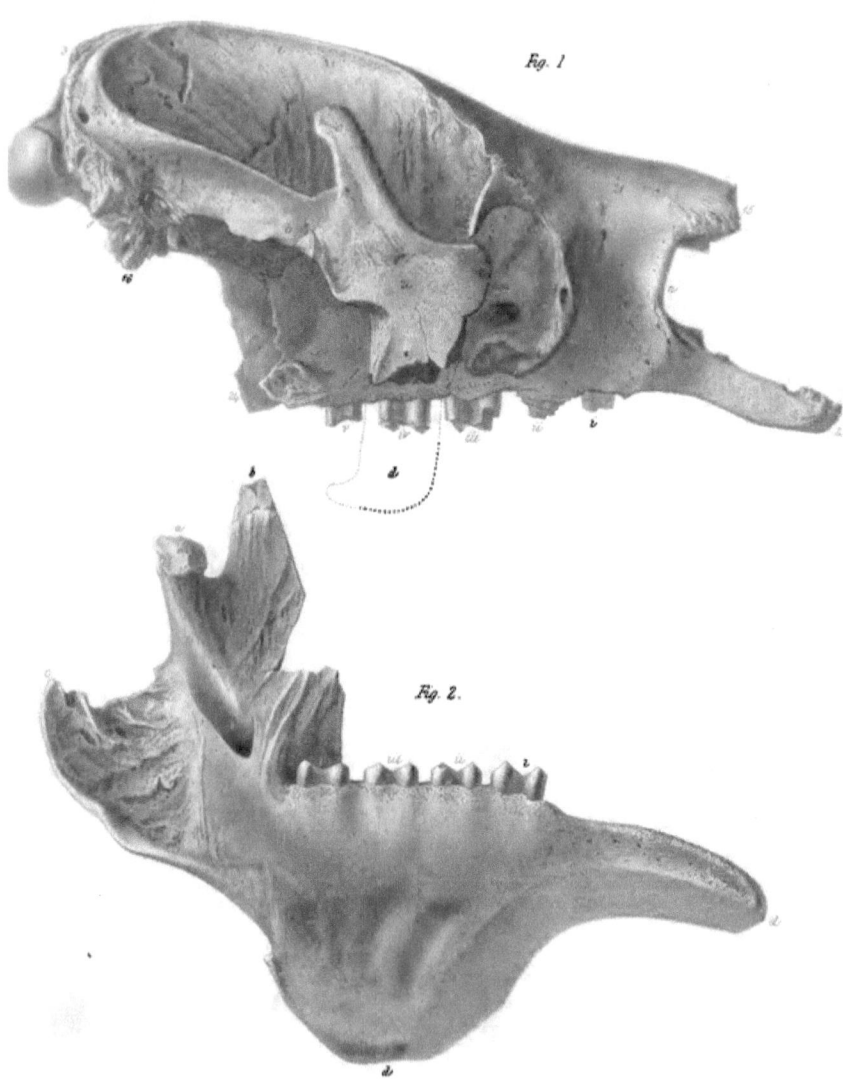

PLATE XII.

One-fourth natural size.

Fig. 1. Side view of the skull of the Megatherium.
Fig. 2. Inner side view of the mandible and of the section of the symphysis.

The letters and ciphers are explained in the text.

PLATE XIII.

One-fourth natural size.

Fig. 1. The opposite and more mutilated side of the same skull as Pl. XII., showing the surface of the temporal fossa.

Fig. 2. **Upper** view of the same skull.

The letters and ciphers are explained in the text.

Fig. 1.

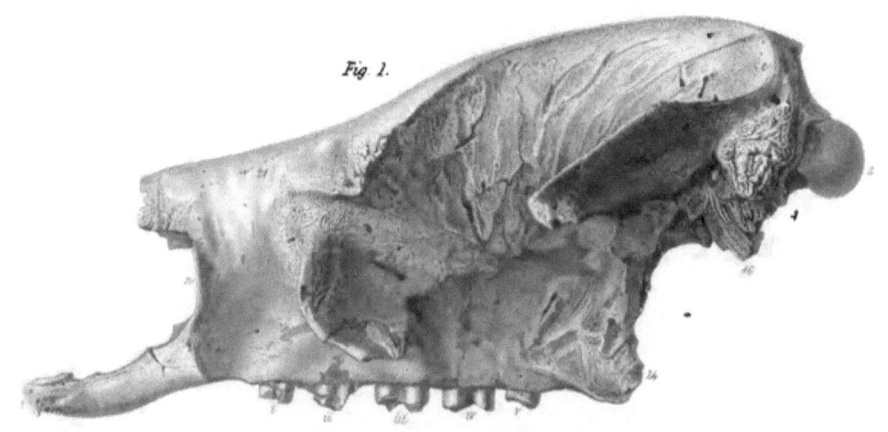

Fig. 2.

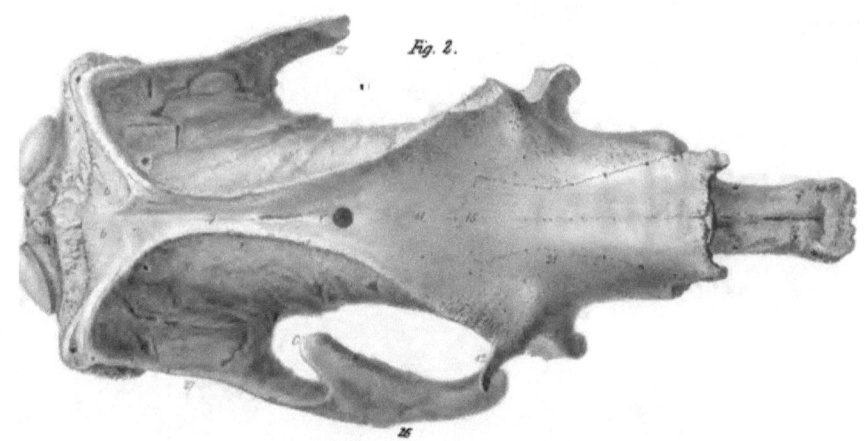

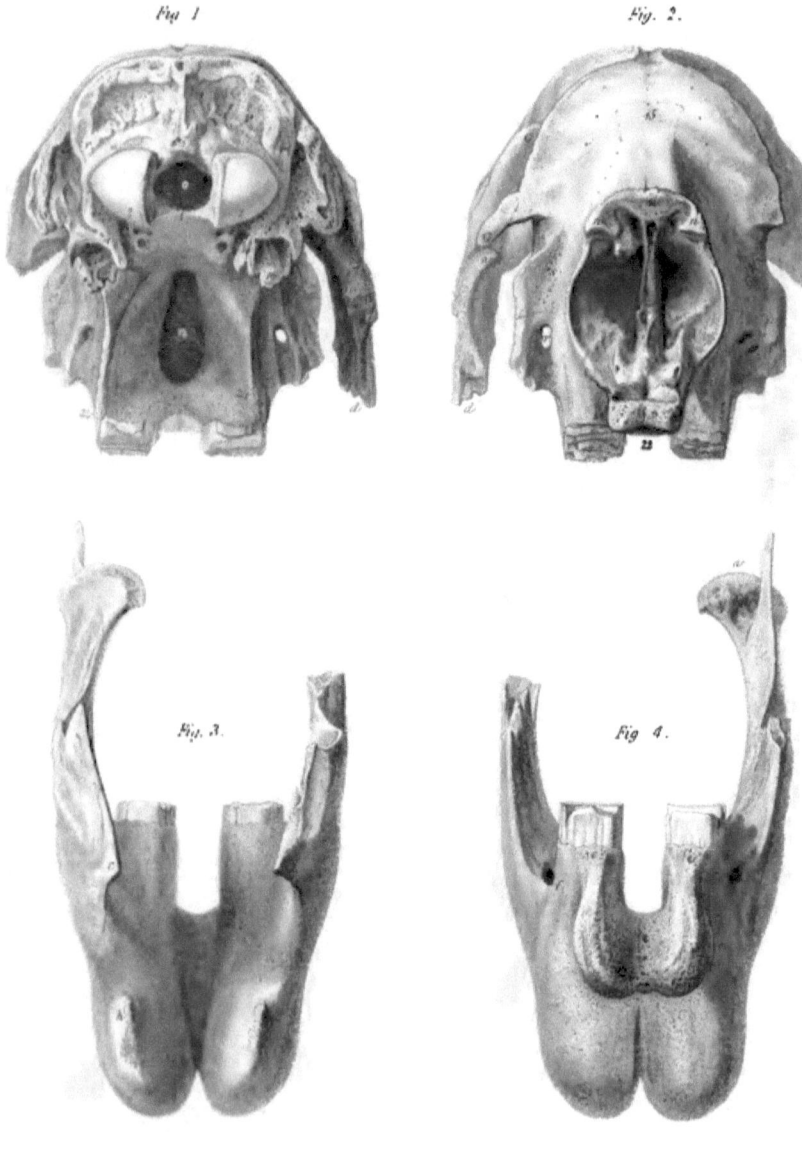

PLATE XIV.

One-fourth natural size.

Fig. 1. Back view of the skull of the Megatherium.
Fig. 2. Front view of the same skull.
Fig. 3. Back view of the mandible.
Fig. 4. Front view of the same mandible.

The letters and ciphers are explained in the text.

PLATE XV.

Natural size.

Base view of the skull, and grinding surface of the upper teeth, of the Megatherium. The letters and ciphers are explained in the text.

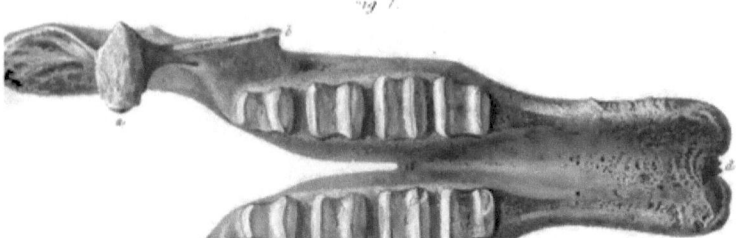

Fig. 1.

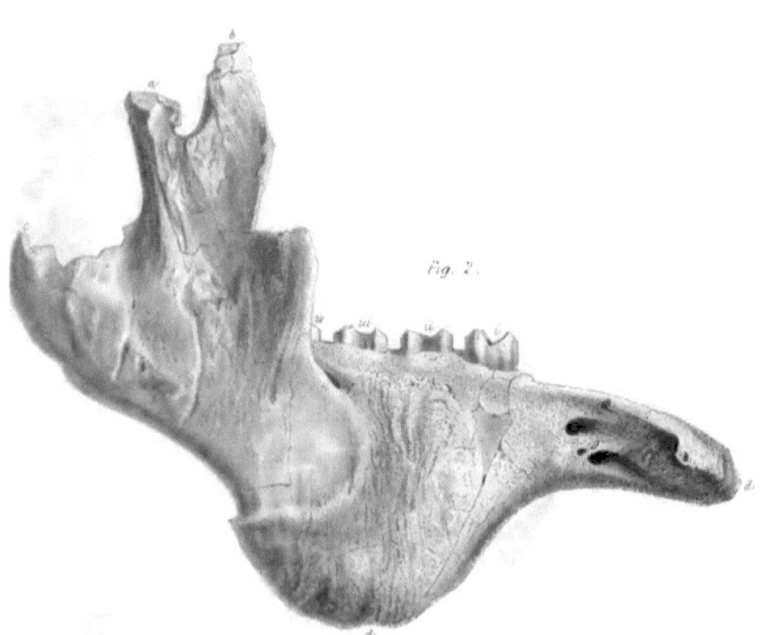

Fig. 2.

PLATE XVI.

One-fourth natural size.

Fig. 1. Upper view of the mandible, and grinding surface of the lower teeth, of the Megatherium.
Fig. 2. Side view of the mandible.

The letters and ciphers are explained in the text.

PLATE XVII.

Fig. 1. The second upper molar tooth—natural size—of the Megatherium.

Fig. 2. Longitudinal section of the alveolar part of the upper jaw and of the five upper molars, *in situ*—three-fourths of the natural size—of the Megatherium.

Fig. 3. Transverse section of a portion of a molar tooth of the Megatherium:—magnified 500 linear diameters.

The letters and ciphers are explained in the text.

Fig. 3.

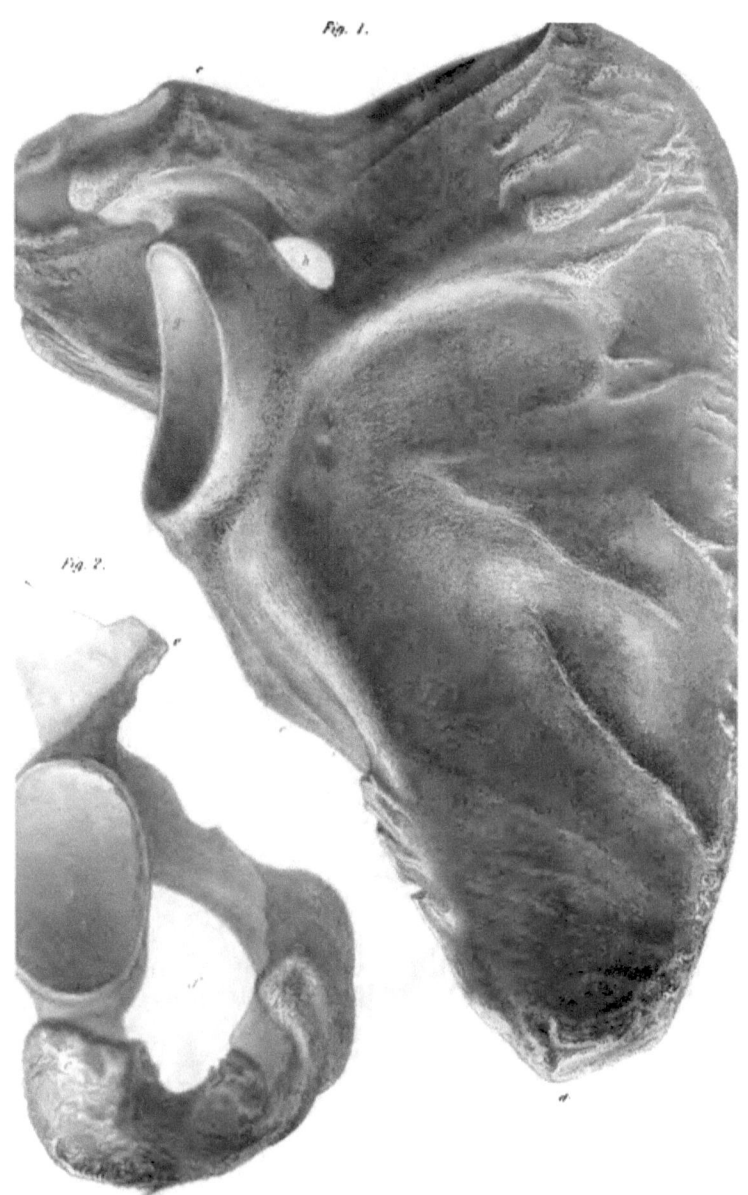

Fig. 1.

Fig. 2.

PLATE XVIII.

One-fourth natural size.

Fig. 1. Inner surface of the scapula.
Fig. 2. Glenoid cavity and acromio-coracoid arch.

PLATE XIX.

One-fourth natural size.

Fig. 1. Under surface of the right clavicle.
Fig. 2. Front view of the left humerus.
Fig. 3. Back view of the left humerus.
Fig. 4. Head, or proximal articular surface of the left humerus.
Fig. 5. Distal articular surface of the left humerus.

PLATE XIX.

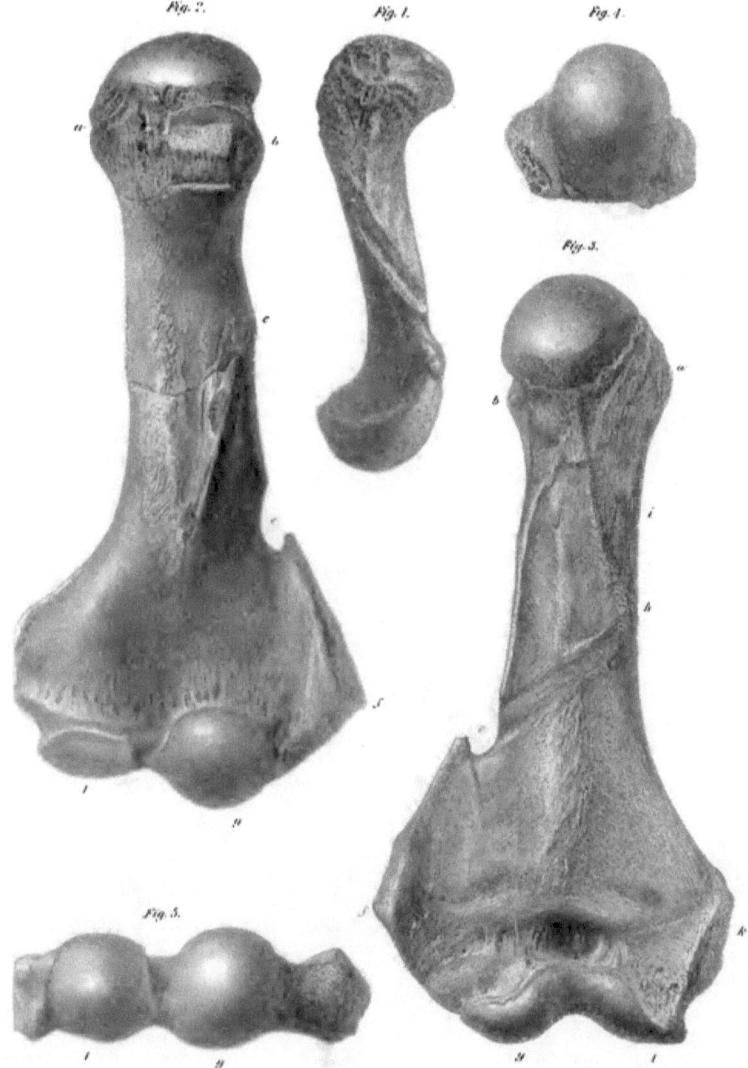

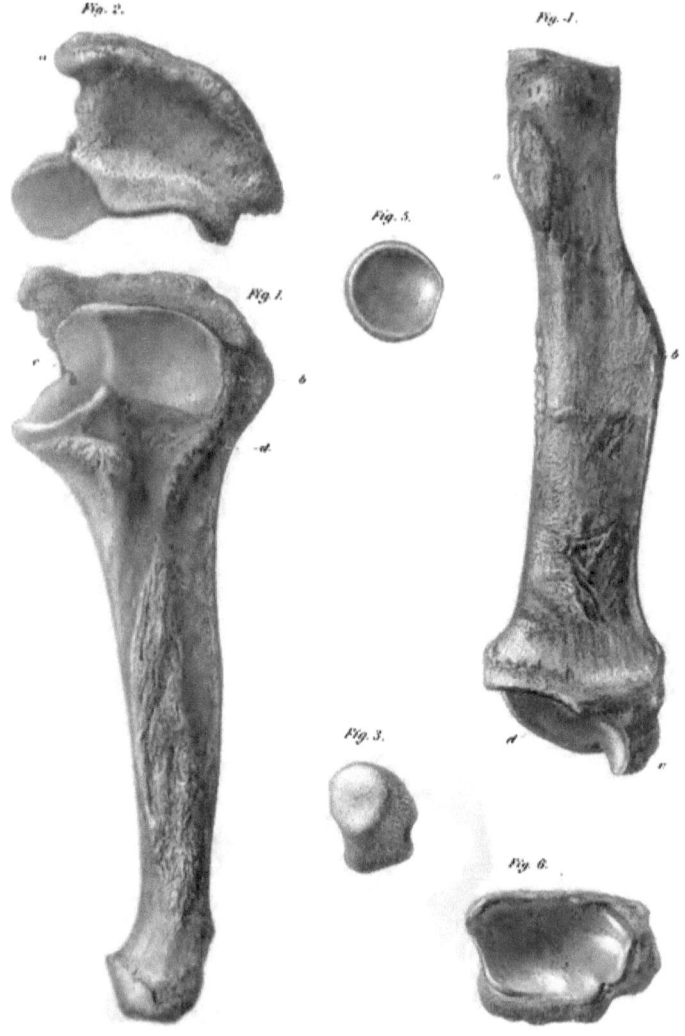

PLATE XX.

One-fourth natural size.

Fig. 1. Antero-radial surface of **the left ulna.**
Fig. 2. Proximal end of the ulna.
Fig. 3. Distal end of the ulna.
Fig. 4. Anterior surface of the left **radius.**
Fig. 5. **Proximal end of** the **radius.**
Fig. 6. **Distal end of the radius.**

PLATE XXI.

Dorsal or upper surface of the bones of the right fore-foot: less than one-half the natural size.

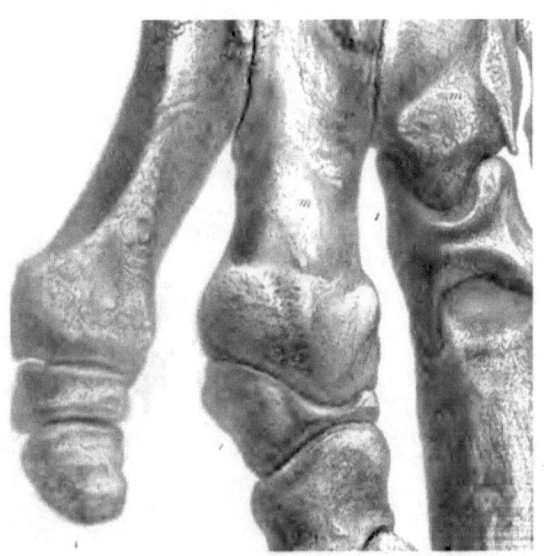

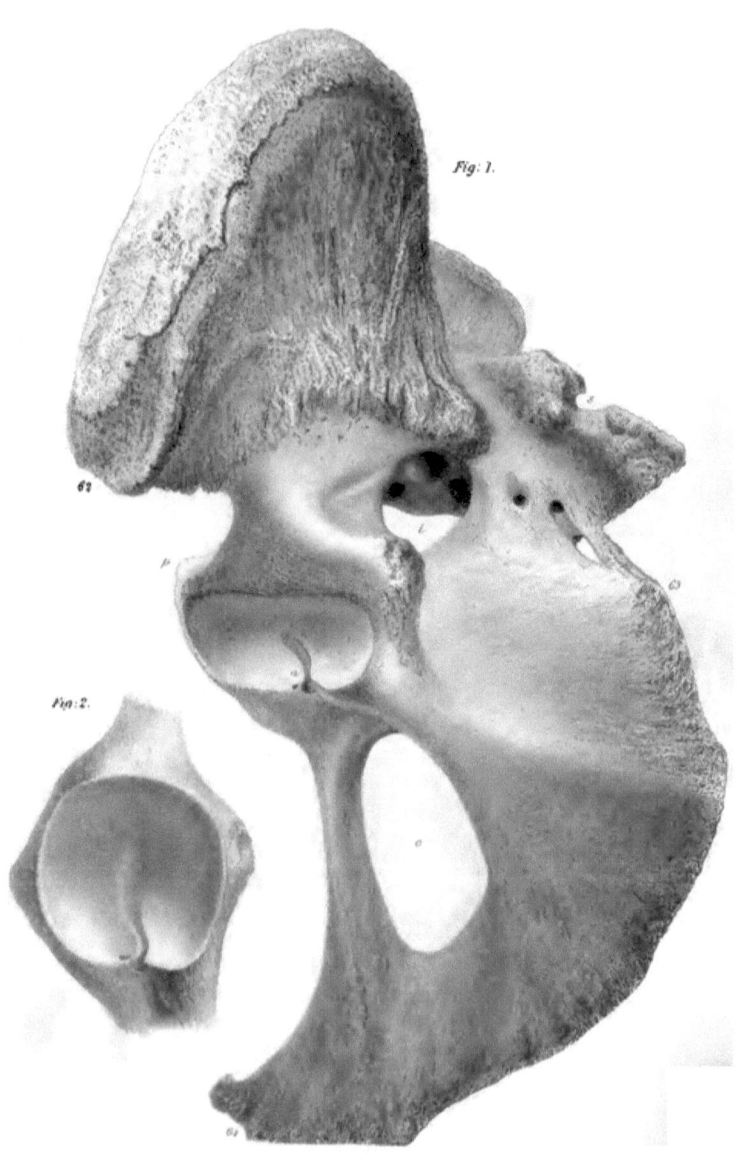

Fig: 1.

Fig: 2.

PLATE XXII.

One-fifth natural size.

Fig. 1. Side view of the pelvis of the Megatherium.
Fig. 2. The acetabulum.

The letters and ciphers are explained in the text.

Drawn, by permission of the President and Council, from the specimen presented by Sir WOODBINE PARISH, K.H., F.R.S., to the Museum of the Royal College of Surgeons of England.

PLATE XXIII.

One-third natural size.

Fig. 1. Back view of the left femur.
Fig. 2. Back view of the left patella.

PLATE XXIII.

Fig: 1.

Fig: 2.

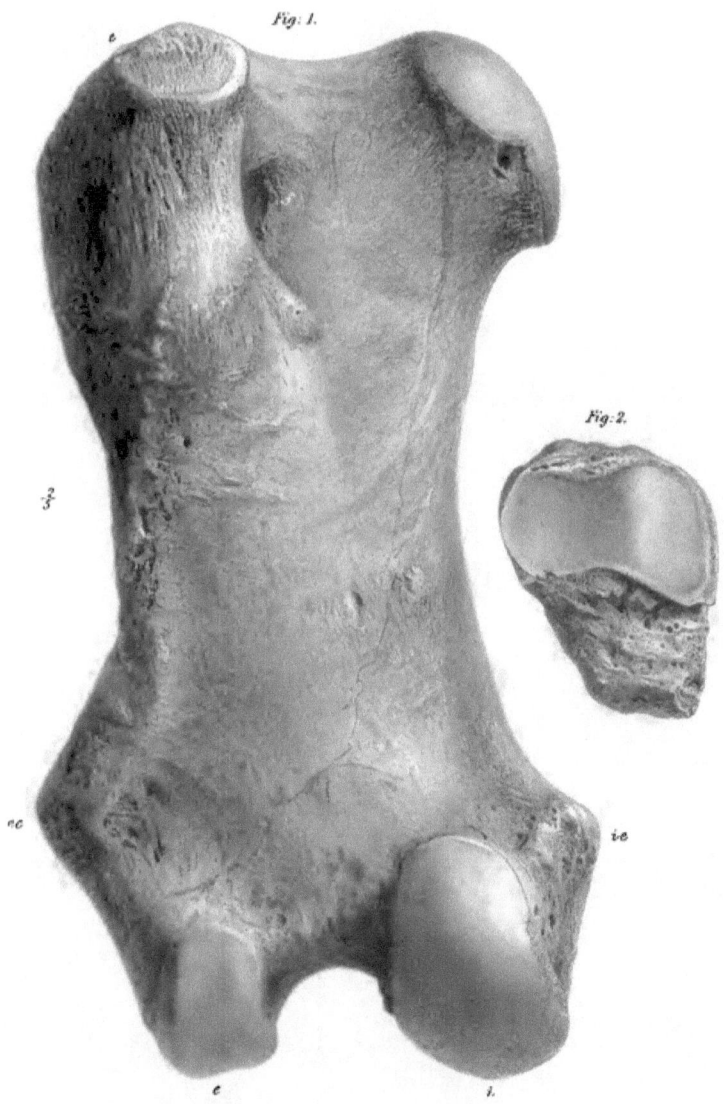

Fig. 1.

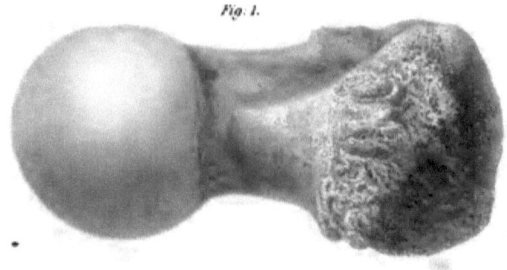

Fig. 2.

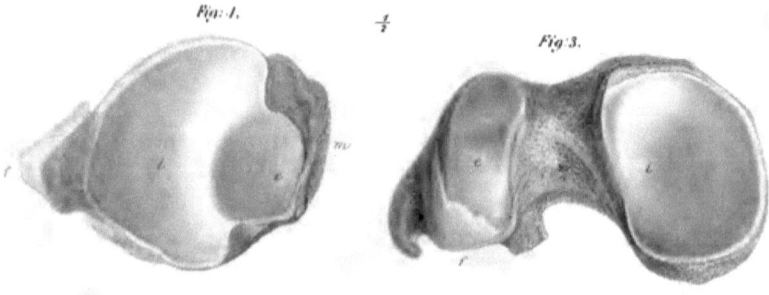

Fig: 4.   Fig: 3.

PLATE XXIV.

One-third the natural size.

Fig. 1. Proximal **end of the femur.**
Fig. 2. Distal **end of the** femur.
Fig. 3. Proximal end of the tibia.
Fig. 4. Distal ends of the anchylosed tibia and fibula.

PLATE XXV.

One-half the natural size.

Fig. 1. Tibial side of the bones of the hind-foot.
Fig. 2. Front view of the two cuneiform and the cuboid bones.
Fig. 3. Front view of the navicular bone, and parts of the astragalus and calcaneum.

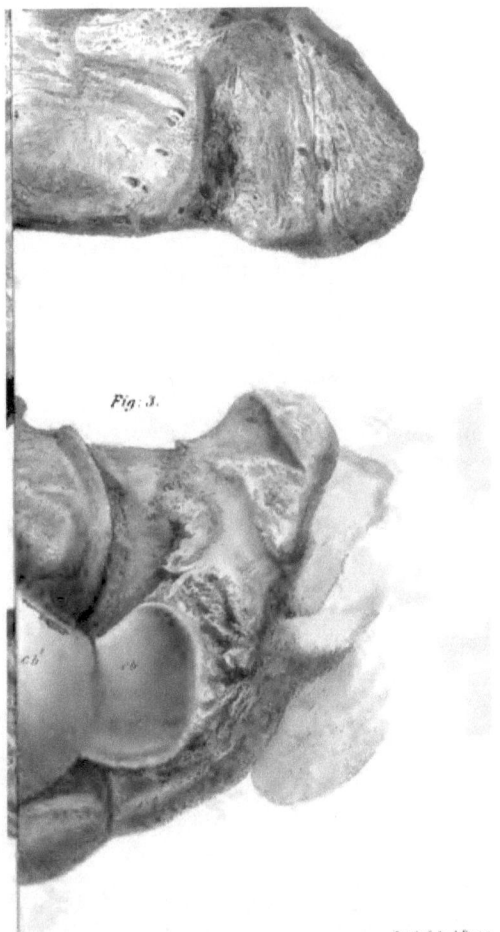

Fig. 3.

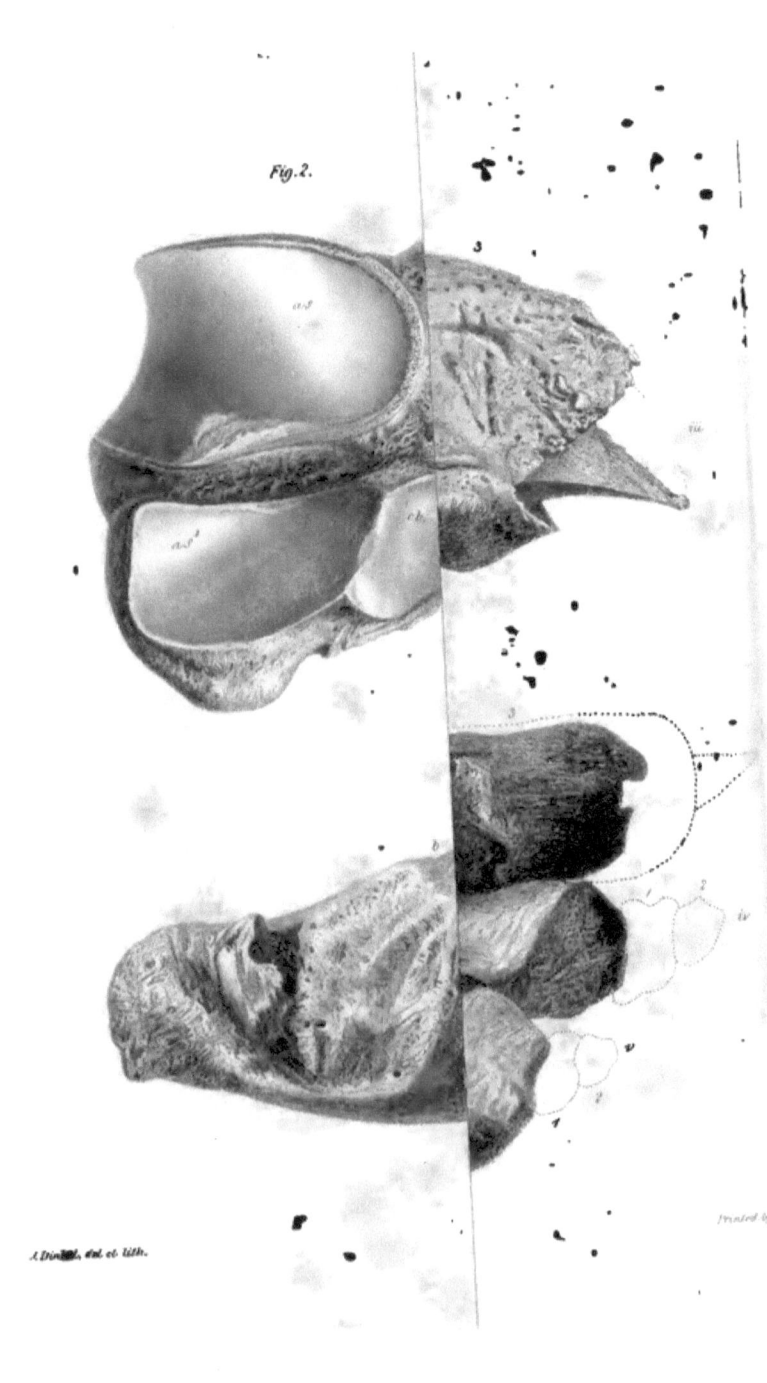

Fig. 2.

PLATE XXVI.

One-half the natural size.

Fig. 1. Upper view of the bones of the hind-foot.
Fig. 2. Front view of the calcaneum.
Fig. 3. Fibular side of the bones of the great unguiculate toe.

PLATE XXVII.

Reduced and oblique front view **of** the skeleton of the *Megatherium americanum*, in the British Museum.

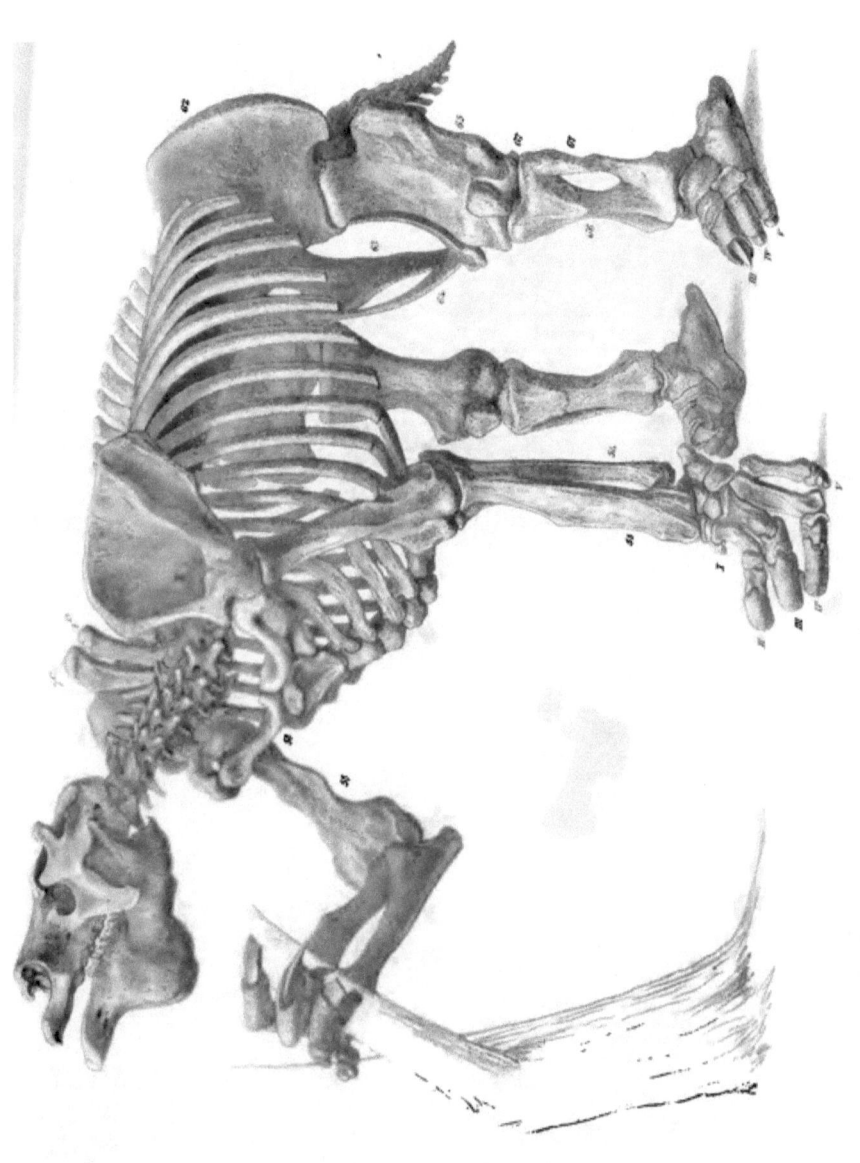

www.ingramcontent.com/pod-product-compliance
Lightning Source LLC
Chambersburg PA
CBHW020306170426
43202CB00008B/517